Praise for Peter Alson's *The Vig*

"A kind of literary Goodfellas."
—*Arena*

D0165302

"Alson shows a knack for revealing character, especially his own. He suggests the ways in which choice, chance and a variety of imperatives fuse, deciding who we are and what we do with our lives."
–Paula Friedman, *The New York Times*

"Painfully perceptive..Alson's bald-faced honesty, which is harder on no one than himself, combined with his considerable talent for reportage, make the book compelling, almost compulsively readable...he has an uncanny ear for dialogue [and has] learned as much about human nature as about point spreads. The result is not only a rare slice of New York life but an unself-seeking chronicle of what happens when a bright future bottoms out."
–Sue Kelly, *Time Out New York*

"Like the court officials in *The Trial*, Alson grows used to, then reliant on, the bad air of his new life...His fellow bookies are hot, bitter and immediately alive."
–Andy Becket, *The Independent*

"With humor and introspection...Alson grapples with his conscience, his fears of arrest, his embarrassment at enjoying 'success' outside the law, and he tells it with candor."
–Digby Diehl, *Playboy*

"This is a record of a young life gone off the rails—not that tragically, though, and often comically...The story, like the writing, is smoothly done."
–Tobias Hill, *The London Observer*

"As wisecracking a gang of rascals as you could meet on a page. And that makes for telling eloquence."
–David Hughes, *London Sunday Mail*

"The book feels, in its telling honesty and loving humor, to have risen unbidden from the unconscious…The proof of his redemptive gamble is the hard evidence of his book."
–Chris Busa, *Provincetown Arts*

"These foolproof lowlife memoirs are cool, funny, and engagingly un-melodramatic…Alson made some dough, had some laughs, saw some life—and, obviously, fixed his writer's block. Where can we apply?"
–Kenneth Wright, *Glasgow Herald*

"I found myself vicariously moved by greed and anxiety…I hail Alson as a talented, witty, hard-boiled sentimentalist."
–Patrick Skene Catling, *London Sunday Times*

"Extremely funny….What is endearing about Alson is that he sees the drama of ordinary life. It is as if he has suddenly woken from a dream to find that he has squandered most of his life. This is a blokes' book and should be read by all men whose midlife crises hit them in their mid-thirties."
–Nick Foulkes, *The Literary Review*

"A racy, reckless memoir. With the style and gusto of a 1990s Damon Runyon, Alson paints a vivid portrait of a shady New York demi-monde…He has a gambler's ear for sharp, witty dialogue and a fanatic's feel for the male bonding peculiar to men bent on risk…He is so good that it is almost a shame that he will no doubt move on to other themes."
–Anthony Holden, *The London Times*

"One of the beauties of Alson's *Confessions* is that he didn't become a bookie to write about it…So it doesn't have any of that glossy coating of so many write-by-numbers books… The reader is as surprised as the author about how things turned out…Eventually, in the writing, he transforms his world and his life into a piece of Americana."
    –Michael Blowen, *The Boston Globe*

"His description of his time in jail, a brief time but a hideously memorable one, brings back a sharp memory of *The Bonfire of the Vanities*, the definitive story of gentility gone awry… This is a fascinating tale of what not to do when you're in desperate need of a buck."
    –Mary Caldwell, *Voice-Tribune*

"Alson's on a roll in a subculture peopled with Day-Glo characters…His unusual story and punchy style make this one a good bet."
    –Pam Lambert, *People*

"An odds-on winner!"
    –Andrew Martens, *Liverpool Daily*

THE VIG

CONFESSIONS OF AN IVY LEAGUE BOOKIE

PETER ALSON

*For Peter and Cusi,
Two of my favorite
writers and people!
Much love,
Pee*

ARBITRARY PRESS
New York

Published by Arbitrary Press

ISBN 000-0-00-000000-0

Typesetting services by BOOKOW.COM

For my mother and father

ACKNOWLEDGMENTS

Thanks are owed to the many who believed even when the evidence did not support it. But especially to:

Julie Rigby for her love and understanding. My agent, Tom Wallace, who recognized the merits of this book when it was mere notes. Jill Kearney, whose interest and excitement really got the ball rolling. Henry Sandow, who inspired me at a real low point in my life and who was willing to put his money where his mouth was. Ralph Rugoff, who knows the story behind the story. And to Hobie.

I'd also like to thank my wonderful editor, Karen Rinaldi, who'd be embarrassed if I got as gushy as I feel. Ann Patty for being smart enough to hire her. My publicist, Hilary Bass. My copy editor, Donna Ryan. And everyone else at Crown who helped make this experience such a positive one.

For support ranging from massive to inchoate, in alphabetical order: Elizabeth Alson, Kate Alson, Kathy Anderson, J.B., April Bernard, Ron Bernstein, Jonathan Black, Steve Chao, B.D., Marion Ettlinger, Steve Fenichell, Jackson Friedman, Anne Glusker, Carol Goodstein, Amy Handelsman, Tony Holden, Paula Kellinger, Matt Klam, Mike Konik, Legs McNeil, Norman and Norris Mailer, David Michaelis, Katherine Mosby, Richard Nalley, Karen Odyniec, Tom Piazza, the Ragdale Foundation, Ivan Solotaroff, Susanna Sonenberg, Gary Taubes, Jeff Tuchman, Al Wasserman, Jane Williams, the Writer's Room, Fred Zollo, and the boys at the office.

And why is gambling worse than any other means of acquiring money?

— Fyodor Dostoyevsky, The Gambler

PROLOGUE

The judge in Brooklyn Criminal Court is gray-haired and hawk-nosed, his mood somewhere between peevish and utterly bored. When Bob, Michael, and I are called before him, he lowers his reading glasses and peers at us like a man not quite sure for a moment what he is looking at. Though we have spent the past twenty-six hours locked up with the worst kinds of thugs and street punks in the depths of the barred concrete hell three floors below, and though we smell so bad that our lawyer visibly recoiled upon meeting us for the first time, on the surface we still look like what we are: young, neatly dressed, well-educated white males. And even the most jaded night-court judge can't help registering a flicker of curiosity. Our lawyer plays this up in making a case that we should be released on our own recognizance and without bail.

"Your Honor," he begins, and starts reeling off our bona fides, pronouncing with particular ostentation the names of the esteemed universities we have attended. I am horrified, and it goes beyond the obvious shame. I am looking at the judge, whose expression is one of perplexity. I am thinking that our lawyer has pushed too hard, that this is going to backfire on us. The judge will be harsh because we have squandered what most of those who parade before him have never had. He will want to punish us for that, not forgive us. And perhaps he will be right. . . .

But of course that isn't what happens. We get a raised eye¬brow, nothing more. He says, "Counselor, am I to understand that times are so tough that Ivy Leaguers are now going into the bookmaking business?" And as the three of us and our lawyer try not to smile, he pounds his gavel and waives our bail.

1

FOR me, it all began in the heat of July, in the season before the great baseball strike, when I let my friend Winnie Katallis talk me into spending a weekend at her Sag Harbor beach house. I'd been back in New York for weeks trying to find a job and having no luck, and the desperate energy I'd marshaled for the task was already dissipating, turning to depression. I was apartment-sitting in Brooklyn Heights in a tiny studio that had neither air-conditioning nor a kitchen. "C'mon," Winnie said to me. "Get away for a couple of days. It'll be good for you. No one's in New York anyway. It's summer. You can network at the beach."

I hesitated, but she was persistent: her twenty-four-year-old son, Michael, was driving out with his girlfriend; I could get a ride.

Four hours later I found myself walking down Tenth Street toward Avenue A, a black Lucas overnight bag slung over my shoulder. It was typical midsummer sweltery New York, the sky a pearlescent smudge over the low-slung tenements, the air as steam-smelly as a hot dog cart. I felt guilty to be escaping. Undeserving.

Michael was waiting at the corner, leaning against a shiny black Saab, talking into a portable phone. I barely recognized him. Last time I'd seen him, he'd been a grungy twenty-year- old in jeans and disintegrating canvas high-top Cons. Now he was clean-shaven and wore a beige linen shirt, chinos, and Topsiders, and his intelligent hazel eyes were framed by Oliver Peoples wire-rims. He had a sleek, bored-looking blonde waiting in the passenger seat.

Seeing me, he clicked off the phone, pushing down the antenna in one swift, smooth motion.

"Hello, Michael," I said, gripping his hand and laughing.

"What's so funny?" he said.

"Nothing. Just that I can't afford bus fare and you're driving this."

"Oh. Yeah," he said without a trace of embarrassment. "Mom said you'd been having kind of a rough time."

"She did?"

"You know, with Chicago and Anna and trying to find a job."

"Yeah, it's just a . . . transitional period. You know." There was no point in getting into true confessions. I was a mess and sure it showed. "Business must be good, huh?" I patted the fender of the Saab.

"Business has been very good, yeah."

If I hadn't known what Michael did for a living, I would have guessed Wall Street. That was the look and the attitude. But Winnie had told me already. He was a bookie.

She was of the opinion that his being a bookie was okay. I saw nothing to make me disagree. After graduating from Brown in 1989, Michael had spent a couple of aimless years living at home, hanging out in bars, and generally making Winnie fret about him. It had been such a relief to her when he started making enough money to get out of her hair that the method hadn't really mattered. Sure, she'd gone through a bit of hand-wringing, wondering if she was a terrible mother or a bad person because her son was making his living illegally. But I don't think deep down she ever thought she was, and I never did either.

I knew about gamblers and gambling. My grandfather on my mother's side had been compulsive, and I had written about world-class poker players and had played some serious poker myself. Professional gamblers—even those who gambled legally—had an outlaw mentality. They were people who had chosen to live their lives outside the socially accepted boundaries, to thumb their noses at the world. I admired them for that.

"You don't worry about getting arrested?" I asked Michael.

"I've been doing it for a year and a half and it hasn't happened yet," he said.

"Isn't it inevitable?"

He shrugged. "I suppose so."

We loaded my bag into the trunk. A Hispanic guy in a wool hat approached us, carrying a couple of tennis rackets.

"You fellas interested in these?"

"How much?" Michael asked.

"Twenty bucks apiece."

"Let me see."

The guy glanced up and down the block quickly and handed Michael one of the rackets. "That's a good racket," he said. "Prince. Practically new. Take a look."

"Your name is Finkel?" I said to the guy, seeing the name written in Magic Marker on one of the white vinyl racket covers.

"Huh?"

"Finkel," I said, pointing to the green letters.

"Oh, yeah. That's me."

Michael smiled. Finkel smiled. Why shouldn't a Puerto Rican Jewish guy be out on the street selling his tennis rackets?

Michael walked around to the passenger side and bent to the window. "What do you think?" he asked his blond companion. "You want one?" He stood up. "All right," he told the guy. "I'll give you twenty for both of them."

"Twenty for both? They're practically new."

"Twenty or nothing." Michael looked at me. I took my cue and slammed the trunk lid shut.

"All right, twenty," Finkel said.

Michael squeezed from his pocket a roll of cash big enough to make Finkel's head recoil. He peeled off a twenty, waited until he had both rackets in hand, then handed the bill over. He had the cool peremptory manner of someone who was young and very sure of himself, very sure of the power of money.

As we pulled away from the curb, he congratulated himself on his acquisition. "That's a pretty good racket, isn't it?" he said, picking up one of the Princes.

"For the guy selling it, anyway," I said.

He laughed. His laugh was not loose or natural. It had a forced, pinched quality. He had a high forehead and small precise features and a small heart-shaped mouth that did not move much when he talked.

"Someone else would have bought them if I hadn't," he said.

"That's true."

We had barely gotten underway when his phone rang. Again I had to suppress my amusement, trying to put together the sullen-faced kid I remembered with the one behind the wheel of this fancy car. It wasn't exactly lost on me that I was thirty-three and unemployed and he was twenty-four with the car, the girl, the unmistakable air of success. Not only that but he was open about what he was doing, not ashamed of it. He didn't bother to hide it from his girlfriend, who had a straight job in the fashion industry but seemed to get a kick out of watching her boyfriend operate.

Michael grew impatient on the phone. "Monkey, let's talk about it when I get back," he said. "These calls cost money." He clicked off.

I said, "Monkey?"

The blond girl, whose name was Lindsay, laughed and said, "Go ahead, Michael. Tell Peter about Monkey."

Michael shrugged. "He's one of the bosses."

"I was going to ask you about that. Who *are* the bosses?"

Michael fiddled with the radio tuner. "Just some guys."

He knew what I meant, but maybe didn't want to say in front of Lindsay. "Are they mob?" I persisted.

Michael laughed his imitation laugh again. "Everybody always thinks that. I work for a bunch of guys who've been in the business for a long time. They're not mob."

"But you're making a lot of money," I said. I would find out later in the weekend that Michael had made $150,000 cash, no taxes, the previous year. "I can't believe the mob wouldn't want some kind of cut."

"They don't." He was adamant. "It doesn't work that way." It was the only time during the entire drive—weekend, for that matter—that Michael held back. Whether he was making phone calls or playing with his sports ticker—a beeper that gave sports results, updated every

4

ten minutes—Michael did it with an exhibitionist flair, as if he was eager for us, his mother included, to see what he did.

I threw questions at him nonstop. Winnie seemed bored by the nuts and bolts of it—except when I asked Michael what he actually did with all the cash he was making and he told me that he'd invested some of it in a restaurant and kept the rest under a floorboard in his apartment. This made Winnie perk up. "Yeah, that restaurant was a great investment." It turned out the place had been a disaster. But when he and I were talking about point spreads and over-unders and the vig, she'd roll her eyes and try and get Lindsay to go for a walk with her or play a game of Boggle or drive to a yard sale somewhere. On the other hand, she loved it when Michael took us all out to an expensive seafood place and paid the check with three crisp hundred-dollar bills. Not once in my entire life had I taken my mother and her friends out to dinner like that.

Earlier that day at the beach, we'd run into a bunch of people Winnie knew, including an editor from *Esquire* who remembered a piece on poker I'd written. He asked me if I was still writing for magazines, and it got me thinking about Michael. He might be a good subject for an article. I didn't say anything—he was standing right by us—but it seemed like a natural. Both Michael and his partner, Bob—the two of them had their own group of bettors for which they received a percentage of their office's weekly take—had gone to Ivy League schools. I was sure I could sell *Esquire* on that angle alone—Ivy League Bookies. As I watched Michael pay the check out of his fat bankroll, my only real questions were Would he cooperate, and Did I want to write it?

Back in Brooklyn after the weekend, I continued to mull it over. I couldn't kid myself that one freelance piece was going to solve anything.

Looking around the small studio at the clutter of suitcases and stacked-up boxes—my furniture was back in Chicago with Anna—I tried once again to assess my future. My failures trailed behind me like tin cans on a string, rattling and distracting, impossible to outrun.

Even if I got an assignment from *Esquire,* the piece would take at least a month to research and write. Then there would be the inevitable delays: editorial inertia, possible rewrites, foot-dragging by the purse-strings people, unforeseen black holes. Even if I was lucky all the way down the line, it would be months before I got paid.

And then where would I be exactly? Back to freelancing? Back to the month-to-month where's-the-check, hustling-my-ass-off, never-really-getting-anywhere routine that I had sworn off with such resolve two years before?

I was at a point where the whole idea of writing was freighted with failure and disappointment. I had spent the winter on the Cape making what I had decided would be my last stab at real writing, trying to fulfill the early promise that now seemed like nothing more than a curse. At Harvard, a short story of mine published in the *Advocate* had been mentioned in the *Crimson* as evidence that literature was alive and well on campus; the same story had been anthologized in a book called *First Flowering*, my name in the index after James Agee's and Conrad Aiken's. Great things were expected of me after that. Ten years later I still had not delivered. The ignominy, the shame, of once again getting first-novel block—how many times had I rationalized previous failures? three? four?—had fairly destroyed what parts of my ego had emerged intact from the disaster with Anna in Chicago. Leaving her, leaving that stolid city, I'd holed up in a large borrowed house in Provincetown for one more try. As always, the whispers of doubt pierced my resolve, and when winter came, settling in with its bone-deep grayness and sense of desolation, I was through.

Limping home to New York, I had it in mind that maybe I'd luck into another editing job the way I had in Chicago. When that didn't happen right away, I started to wonder if I really wanted it to, if perhaps I shouldn't just abandon the world of words entirely, even though it was the only world I knew.

I still hadn't decided what to do when my mom, an ex-book editor, made her weekly check-in call from her summer beach house on the Cape. Just to have something to talk about, since there was nothing to report on the job front, I tried out the Michael idea on her.

"Do you really want to do another gambling story?" she asked.

"I don't know, Ma. I'm just putting stuff out there. I was thinking maybe I could do a whole book of pieces in the same vein. You know, *Jobs You Can't Tell Your Mom About.*"

She laughed and said, "I thought you said Winnie knows about Michael being a bookie."

"Yeah. She does."

"So you just mean jobs you can't tell *your* mom about?"

It was my turn to laugh, albeit with a pang, for there was a part of me that was jealous of Michael, that suspected he could do this thing, which I thought of as romantic and dark, and not be afflicted by guilt. My own psyche did not let me off the hook so easily. Although I'd rebelled against my mother's squareness at times, I was never free of the guilty price those rebellions exacted, the sense that I knew better, was better, could be better. I dashed off a query letter to the *Esquire* editor.

A week later I was still waiting to hear from him when the subject of my proposal called me.

2

ON Anna's birthday I rented a video called *Hot and Nasty*, whacked off to it a couple of times, and spent the rest of the night in a depressed stupor of self-pity and loathing, watching talk shows on TV and eating Sun Chips until I could feel grease seeping up from under my skin.

I knew what a big deal birthdays were for Anna. But what could I do? She'd apparently made up her mind about us. Just as I'd made up mine nine months before. Minds could change, of course. Mine had. But then, if I were to tell my side of the tale, it would begin, "I didn't leave because I stopped loving you. ..." The fact was that the months had softened the memory of the pain we'd caused each other. Distance, time, and loneliness had rekindled hope.

When I had started calling Anna from Provincetown, which happened one freezing, wind-whistling night in February, suggesting that I might like to come back (the calls increasing in frequency as the pace of my work fell off), she, rightly, demanded to know what had changed for me to be talking like that. It wasn't as if I'd gone off and fought a war or climbed a mountain or won my fortune (though in a sense I'd been trying to do all three); a return suggested no triumph or epiphany. I was at sea. That was as apparent to her as to me.

In fact, the only thing that gave me any credibility with Anna at all, that made her think there might be something beyond desperation involved in my efforts to get her to try again, was her utter disbelief that she "could be the organizing principle in anybody's life." But finally even that wasn't enough for her to say Come. She was seeing someone

—"It's not like with us, but I'm not going to tell you it doesn't matter, either"—and she'd resolved to move forward, not backward. Eventually, as I kept up my campaign, she suggested that it might be a good idea for us not to talk for a while. Which had been the state of things for the past month.

Now, couch-bound in Brooklyn on her birthday, staring at a Mexico-shaped water stain on the ceiling, I could hear her voice, the way she said my name, her mouth pressed up against my ear . . . and I thought of her seven-year-old son, Nathaniel, who on the day I left would not look up at me from the toys in his bathtub, this boy whom I loved and who had loved me back. . . .

When the phone rang it was like a shot of adrenaline to my heart. I sprang up, hitting the mute button on the TV, letting the phone ring twice more, foolish heart beating crazily.

It was Michael Katallis.

"Did I get you at a bad time?" he asked.

I stifled a laugh, looking at the wadded-up Kleenex on the floor by the wastebasket (I'd missed), at the nearly empty bag of Sun Chips, the mute tableau on the tube of Jenny Jones and a group of women who loved her too much. I told him no, it was a fine time.

"Are you still trying to find work?"

It didn't take much at that point to figure out what he was calling about.

"You seemed interested in the business," he said.

"I guess I was, yeah."

"I thought you might want to try it."

"You mean working with you?"

"Yeah."

A sniff of amusement escaped me. Michael said nothing, but I could sense his impatience. He didn't consider this a frivolous call.

"I'd need to know more," I said.

"There isn't much to tell. It's three hours a day. Thirty bucks a shift while you train, fifty-five if it works out. We'd have to see how it goes."

"Fifty-five bucks? That's it?"

"In cash."

"And this would start when?"

"Tomorrow."

"Tomorrow?" Despite myself, I now did laugh out loud.

"Is there something funny about that?"

"No, it just doesn't sound as if I have much time to think about it."

"Well, we need someone. So if you're not interested I have to know."

"Does your mom know this is why you wanted to get in touch with me?"

"I didn't tell her. Why? Do you care?"

"No, it's not that. . .." But what was it? Michael was her son. She knew about him. If that was okay, why was I afraid she'd think badly of me? "I just need to give it some thought," I said.

3

THE office was in a tenement building next to a bodega on St. Marks. I stepped around a grunge punk who was using the dirty window of the door as a shaving mirror, scraping the scalp on either side of his spiky Mohawk with a dry razor. Three sharp rings to the buzzer, per instructions, and I was in. Five flights up beyond the urine-scented vestibule, a door opened and a pudgy-faced blond kid in a Malcolm X hat led me into a gloomy, cluttered apartment filled with Salvation Army furniture, dusty books, peeling paint, and a sour haze of cigarette and b.o. aroma.

Three men sat at a large oval table lit by a low-hanging fluorescent fixture. The blinds in the apartment were drawn. In one of the darkened windows an air conditioner labored noisily and without much apparent effect.

Only Michael looked up at me. "You're late," he said.

A clock on the wall behind him showed ten after five.

I thought he was joking and smiled. He didn't smile back. "Late on your first day," he said, shaking his head. I still couldn't tell if he was serious or not. "Lucky for you it's going to be slow anyway." He introduced me to the blond kid who'd opened the door, Spanky; to Pat, a florid-faced Irishman with a shock of white hair; and to Bob, who had the broad shoulders and the thick neck of a linebacker and wore a green Lacoste shirt over a white T-shirt.

Spanky and Pat nodded noncommittally. Bob said, "How ya doin', Pete?"

"All right. Should I just watch or what?"

"Sit over there," Michael commanded, gesturing to a seat at the end of the table. One of the six phones rang, and Pat, the Irishman, picked it up. "Yeah," he said brusquely. "Not open yet. Call back in fifteen minutes."

I sat where I was, listening to the four of them converse. Somebody's "bottom figure" was off by three thousand; an agent was asking for a higher commission; someone called Benny Cadillac had "a new sheet" for the office that was "fifty percent, fifty percent—a Dutch treat." I had no idea what any of it meant. The phones continued to ring until Bob said, "Fuck it. Take 'em off the hooks. Let 'em wait," and the other three men took the receivers off the hooks.

Pat turned to look at me. He had a mottled pink face—a real Irish busted-capillary complexion—and eyes that were acid-washed blue. "I hear you're a Harvard man."

I shrugged.

"You must have fucked up pretty bad to be here, huh?"

"Hey," I said, nodding at Michael and Bob, "these guys went to Ivy League schools, too."

"Yeah, I already know about these bums. What's your story?"

"You mean where did I go wrong?"

"Everybody's got their story."

"What's yours?" I said.

"My story?" He laughed. "You want to know my story?" He looked from Michael to Bob. They were both stifling smirks. "I was on the executive board of two different companies."

"You're kidding. Two companies?"

"In the rag trade. You ever hear of. . ." He ticked off the names of a couple of well-known clothing manufacturers. "Well, I ran those."

He was serious. Michael and Bob said nothing.

"So what happened?" I asked.

"What happens. One of them went under. The other made an executive change."

"Just like that?" I said. "Was there a reason?"

He shrugged. "A million reasons. And not one of them my fault. Now look at me. I'm sitting here with you fucks. I've got alimony, car payments, and my kid had to take out a loan for college. He'll be paying $292 a month for the rest of his life." Pat paused to let that sink in for a moment.

"Not that he's going to pay it," he continued. "I will. I just haven't told him yet."

"I had to pay off my own loan," Michael said.

"Yeah, and look what your college education got you. A fucking Ph .D in collecting the vig."

Before the office opened up, Michael made calls to several bookies around town to get their opening lines. Later I learned that the lines originated out of the Stardust Hotel in Las Vegas, put out by a well-known oddsmaker named Roxy Roxborough who was paid by the casinos to post a number that would anticipate the way the public would bet. Once he had a few sets of numbers, Michael averaged them out, or shaded them to sides he favored—that is, made it more appealing for bettors to take teams he thought would lose by giving them a cheaper price. When he was done, he said, "Ready?" and began reading off the spread: "Phillies two-ten, Astros sixty cents, Mets twelve cents . . ." while the rest of the crew worked their pencils on their sheets.

I asked for an explanation of the numbers, and Michael said, "I thought you knew this shit."

"I know point spreads on football. Not this."

"Baseball works with a dime line and a twenty-cent line," Spanky piped in.

"You want to explain it to him?" Michael said.

"Sure. See, Pete, let's say you got the Astros, who are sixty-cent favorites over the Padres. That means that if you want to bet the Astros you have to pay a hundred and sixty to win a hundred. You understand that?"

"I think so."

"And if you bet the dog, the Padres, you're betting a hundred to win a hundred and fifty. That ten-dollar difference is the juice. You know about juice, right?"

"Sure." The juice, or the vigorish or the vig or the eleven-to-ten, was the ten percent commission that bookies collected on all losing bets. It was the profit margin, assuming they could get even action on both sides. "Basically, you're trying to get the same amount of action on both sides, right?" I asked.

"Most of the time."

"And when you get too much on one side, you lay off?"

"Hey, he knows about laying off," Pat said.

Laying off was what they called it when a bookie who was getting too much money on a particular team called another bookie and made a bet on that same team, essentially giving away some of the business but at the cost of a five percent commission.

"Yeah, sometimes we'll go to the outs," Michael said. "Outs," I learned, were bookies who took other bookies' layoff bets. "It depends."

"On what?"

"On whether we're on the right side or not. But you don't need to know this stuff right now."

"What do I need to know?"

"Just how to give lines to customers and write tickets."

"So how would it work with the Mets line?" Spanky asked me.

The Mets were twelve-cent favorites over the Cubs. "Let's see, the Mets would be a hundred and twelve dollars to win a hundred."

"That's right! Very good! The kid's a genius," Bob said.

"And the Cubs?" Spanky asked.

"Ah, they'd be a hundred and two to win a hundred?"

Spanky looked at me expectantly. I saw Bob and Michael raise their eyebrows. "It'd be the other way around, wouldn't it?" Spanky said. "A hundred to win a hundred and two. Right?"

"Oh, yeah. Right."

"Harvard," Pat grumbled.

"All right, that's enough," Michael said. "It's time to open."

"Let's rock and roll," Bob said.

The minute the phones were back on their hooks they began to ring. Each of the six telephones sat on top of a tape recorder. All conversations were recorded, protection in case a customer decided to dispute a bet. On one wall six jacks were lined up. The lines worked on the hunt system—incoming calls kept searching until they found an open line. The whole setup looked way too elaborate for this dump of an apartment.

"What did the guy who installed the jacks think was going on?" I asked.

"We told him we were starting a travel agency," Michael said. The others laughed.

Over the next hour I watched as Michael, Pat, Bob, and Spanky fielded a steady stream of calls from around the country. They scribbled orders furiously on triple-sheeted betting slips, then flicked them into an empty cigar box in the middle of the table. Michael was charting—keeping a running account of how much was being bet on both sides of each game, so that he could make adjustments in the line. The lines were everything, I was informed, and it was crucial to keep adjusting them up or down to encourage equal betting on both sides. If too much money was coming in on one side, the price was simply adjusted in favor of the other side until it reached a level at which bettors would find it attractive to take the side we needed.

"Sometimes the balance gets way out of whack," Michael said, "and we'll lay off some of it to another office. But we don't always try to balance our books. Our feeling is that in the long run the vig will take care of us. A guy's gotta win fifty-two point five percent of his bets just to break even against us. The key, really, is volume. If we get enough, we can't help but make a profit."

During one lull in the action I learned that the apartment we were in belonged to a guy named Krause, who was asleep in the next room. I wondered what kind of person would allow such an invasion into his home even if he was getting, as I was told Krause was, his phone and utility bills taken care of and a couple of hundred bucks a week on top of that. Forget the fact that the living room was a pigsty—crooked paintings, a film of dirt on the walls, a flaking ceiling. Forget even that the tired-looking furniture had been shoved into corners to make room for the big fluorescent-lit worktable and that brown shopping bags were scattered around, piled high with garbage and cigarette droppings. How could anyone tolerate the daily violation of his privacy?

At ten to eight Spanky started sorting through the pile of tickets that had accumulated in the cigar box during the session, arranging them in some kind of order. When he finished, he spit into his hands, rubbed them together briskly, and with tremendous speed and dexterity, began separating the white top copies from the yellow bottom copies. The pink bottom copies had been torn off and "ducked" into a hiding place at the time the tickets were written.

Michael said, "You don't have to stick around while we close up, Pete."

I took this as a cue to leave rather than an offer, and I got up.

"Tomorrow's going to be busy, but why don't you come in Thursday morning? It'll be slow enough so you can learn some stuff—that is, if you still think you want to."

"Yeah," I said. "I think I do."

I walked to the subway station through the hot, crowded, still-light streets of the East Village, which seemed different to me now than they had a few hours before. I felt as if I had been walking around with blinders on, oblivious to the secret doings that I now suspected were taking place inside every run-down tenement, behind every facade and storefront. I was excited by what I had seen in Krause's dark apartment.

Waiting for the IRT back to Brooklyn, I couldn't help wondering what would happen if I started making the kind of money that Michael was making. What would Anna think? Would she be horrified? Amused?

I spun out a whole pipe dream just like that, me with cash buried under my floorboards, a fancy car, a plush crib. It was a nice fantasy. But dangerous because it got me thinking about Anna. It made me want to know what would happen, how she would feel if I could actually offer her something more than love.

Back in Brooklyn, I climbed the five creaky flights to my studio, lugging an armful of magazines I had purchased at the Clark Street station in an effort to bring myself back to earth. It was hot in the apartment, and I grabbed a Coke out of the refrigerator and sat in the red velvet armchair by the open window, happy for the slight breeze off the river. Still breathing hard from the climb, I paged through *Men's*

Health and *Men's Journal* and *Fitness and Health,* jotting down the names of editors to contact, along with mailing addresses, telling myself what I needed to do was keep up the job hunt, not relax it or get diverted. Working for Michael was a short-term thing until I could find other work. That was all. Fate had thrown it into my lap, and I had followed up because it was expedient, a way to relieve some of the money pressure, the anxiety. And it gave me something to do. . . .

4

O N Thursday, when I arrived at the office, Spanky handed me a list
of names with some columns of figures beside them. Michael and
Bob had hero sandwiches spread out before them, bits of lettuce and
tomato spilling onto the white waxy wrapping paper. The two of them
chewed noisily, nodding when Spanky said, "He should learn this shit,
right?"

With his high-pitched wise-guy voice and the scraggly blond hair
spilling out from under his X cap, Spanky was a nineties version of a
Dead End Kid. He had been the most junior clerk until my arrival, and
he seemed to take pleasure in being able to show me the ropes. "These
are the weekly figures," he said. "Study 'em. Familiarize yourself with
the names."

The names were grouped in bunches, and next to each group Spanky
wrote another name, using a felt-tipped pen. "This is the name of the
agent—the sheet name. Each cluster of bettors is part of a separate
sheet represented by an agent." He pointed to one group of names:
Dodge, Jaguar, Pinto, Camaro, Ford, and so on. "Like, the agent for
this group is Tranny. When Dodge calls, he'll say, 'This is Dodge for
Tranny.' "

"What do the agents do?" I asked.

Spanky looked at Michael, who nodded at him. Spanky explained:
"Each group of these bettors has an agent. He's the guy who's responsible for them. They pay him and then he pays us, or vice versa."

"What if they don't pay him?"

"If a guy doesn't pay, it doesn't matter. The agent still has to pay us."

"Yeah, but what if the agent doesn't pay?"

Spanky ignored the question, forging onward. "Okay, when you take a bet, you write both names down at the top of the slip. So a typical bet might go, 'Dodge for Tranny, Mets plus fifty-three, for a dollar.' Which means?"

I shook my head.

"That Dodge is putting a hundred dollars on the Mets, who are fifty-three-cent underdogs. So if he wins, he gets how much? "

"Uh …"

"A hundred and fifty-three for his hundred-dollar bet."

"Right."

"And if he bets against the Mets?"

"If he's playing the favorite, then he's paying a hundred and, uh…"

"Sixty-three. Remember? The ten-dollar difference is the juice."

I looked over his shoulder as he began writing something on a betting slip. "For the suckers, though," he said, sliding the betting slip over, "we give 'em a different line."

"How do I know who the suckers are?"

"You'll learn after a while."

I picked up the slip of paper. He had written: 0-10 = 5-6; 10-20 - 5.5-6.5; 20-30 = 6-7; 30-40 - 6.5-7.5; 40-50 = 7-8; 50-60 = 7.5—8.5; 60-70 - 8-9; 70-80 = 8.5-9.5; 80-90 = 9-10; over 90 it becomes a 30-cent line.

My head was beginning to spin.

"Keep this until you can make the conversions yourself," Spanky said. "When a sucker calls, let's say we're using thirteen cents on the Dodgers. You look at this chart, it becomes five and a half, six and a half. Five and a half is actually a hundred and ten, and six and a half is actually a hundred and thirty. Which means that the sucker is betting a hundred to win a hundred and ten when he takes the dog, and laying a hundred and thirty to win a hundred when he takes the favorite. So he's paying twenty percent vig instead of ten. It seems more complicated than it really is."

I wanted to say "I'll bet," but I refrained.

Spanky wrote out another key for me on a separate betting slip. This one translated the shorthand terms for different units of money. In the language of gamblers a dollar equaled $100 and fifty cents equaled $50, but just to confuse things a nickel was $500, a dime was $1,000, and the times sign—an X—was five dollars, as in 20 X equals $100.

After the session I went back to my Brooklyn cubbyhole. There was no mail of any interest, and the only message was from my mom. Sad. I pulled out the stapled pages that Spanky had given me and looked them over. There were eight columns of numbers next to each name— one number for each day of the week and a total at the end. Six pages of names. I counted three hundred. In each column, there was a positive or negative amount. There seemed to be more negative amounts than positive. Wingnut had apparently lost $8,000 for the week, $5,000 of it in one night. Scarecrow had lost $12,500. And a player named Meat had lost a grand total of $23,220. Holy shit. These were not kidding-around numbers.

I put the sheet down and studied the scraps of paper on which Spanky had written out the various keys and explanations. Dizzying. I took a break and returned my mother's call.

My mother's approach to my various problems, both monetary and otherwise, was definitely of the Western medicine school and entailed dealing with symptoms, not root causes. Guilt, disappointment, and anger were things we rarely talked about directly.

"There were a couple of want ads in this Sunday's *Times*" she told me right off. "Did you see them?"

"Yeah, I circled a few."

"Any other leads?"

"Not really."

"What about your friend at *Time?*"

"He says they're cutting back. Definitely not hiring."

"Have you given any more thought to freelancing?"

"Uh-huh. But I talked to a couple of editors I know, and they both said they're not making new assignments right now; they're working off of inventory. The magazines that aren't folding are cutting back. It's grim."

"What about *Esquire?* Did you decide against writing a piece about Winnie's son?"

"No, I queried the guy. I haven't heard back from him. But you know how that is. It's why I don't want to get back into freelancing."

"You're going to have to do something, Pete."

"I understand that," I said, and in a flash of anger nearly added, "I am doing something, Ma. I'm working for a bookie," just to see what she would say.

5

BACK at the office on Monday, I sat and watched. I still didn't understand what was being said half the time. It was like a club or a secret society, with its own special language and codes.

At one point, Pat, the Irishman, took a bet and with some urgency asked whoever was on the other end of the line, "Is that them?" Slamming the receiver down, he said, "That's them," and suddenly there was a frenzy of activity. The betting line was moved, and Michael and Bob punched at the programmed buttons on their phones, trying to get down bets with other bookies at the old price. I felt like I was on the trading floor at Wall Street watching a run on sugar or wheat. The office was trying to buy cheap, sell expensive. That much I could grasp. There was a bristling, uptight energy in the room, unique and unmistakable. The energy of greed.

These frenetic outbursts, which happened several times during the course of a three-hour shift, were instigated, I gathered, by a few different bettors. Pat and the others variously referred to them as "the wise guys," "the sharp guys," and "the computer." Almost immediately after a call came from one of them, "the followers" (the bettors who took their cue from the wise guys) would descend, ringing the phones off the hook, trying to get a bet in on the wise guys' side before the price got too high. During these brief sieges, I was ignored. And afterward it was as if I remained invisible, Pat and Bob and Michael and Spanky talking over me and through me, discussing what had happened.

"These sharp guys," I said, when I was sick of being ignored. "Do they always win?"

"That's why they're sharp," Pat said, showing all his teeth in a smile.

"No, I mean what's their percentage?"

"It's very good," Pat said. "They had a great year last year. Football season, they just destroyed the lines. It doesn't always go their way. But in general? Let me have Danny D.'s bank account, I'd be a very happy man."

"Danny D.?" I said. "The poker player?"

"Yeah. You know him?" Pat was surprised.

"I interviewed him once for a poker piece I did. His father's a writer."

"Yeah, D. Him and Billy Walters are the best sports bettors in the country."

Outside in the bright sunlight of St. Marks after one of these early training sessions, Michael said, "So how does it feel like it's going?"

"All right. There's more to it than I expected."

"You understand the difference between the dime line and the twenty-cent line?"

"Yeah, it's beginning to sink in."

"The important thing in the beginning is just to get the basics. How to give a line, how to write a ticket. Mistakes can be very expensive."

"I can see that."

"Maybe you don't have the head for this," he said. "Not everyone does."

I felt myself beginning to heat up, an "Oh, yeah?" orneriness surging through me even though I knew he was probably right.

6

NEAR the end of that week I got my first taste of working a phone. I was nervous. Pat, Bob, Spanky, and Michael watched and listened, which made things worse. When I actually took a bet from someone, the experience was disconcerting.

"Let me have the Cubs plus the thirty-two cents for a dime," the voice that had identified itself as Pig for Jack said.

I wrote down what he said and hung up. But even with the bet there in black and white, I lacked confidence that I'd done it right. It was as if I had taken down instructions in a foreign language and had no real certainty that my notations meant what they were supposed to mean.

Spanky said, "You didn't give him a read-back. You have to give a read-back after you take a bet."

I looked hopelessly at the phone in its cradle and nodded, even less confident now than I had been a moment before. It wasn't a particularly good way to feel, considering that the transaction had involved a bet of a thousand dollars. What if I'd gotten it wrong?

"Don't worry," Bob told me. "You fuck up, you just pay for it, that's all."

I looked around and no one was smiling.

"Hey, we all went through it," Michael said. "Spanky, what did it cost you? Four dimes?"

"Something like that," Spanky said.

"You mean I have to pay if I make a mistake?"

"You don't think that's fair?"

It took me another minute to see through the deadpans. It was part of some kind of macho initiation process, and I wasn't accustomed to it. Despite the fact that Michael had brought me in, it was clear I wasn't going to get off easy. Beneath the surface hazing, I suspected their suspicions about me ran deeper, like Who the fuck are you? And what are you really doing here?

Good questions, both.

Given my insecurity, it was somewhat of a surprise when I picked up on one of those early phone calls and a gruff voice said, "Hi, Pete. How's it going?" It turned out it was the Monkey, who knew who I was even though I didn't know him.

I handed the phone to Bob. "Hey, boss," Bob said. "How's Vegas? You winning? Yeah? You fucked a girl? Without paying her money? Bullshit. I don't believe it." Bob laughed and put his hand over the mouthpiece. "He says his wife paid her." Everyone laughed. "Listen to me," Bob said. "Don't come back here. We're making lots of money without you. I gotta go, good-bye." He hung up abruptly. "Wait'll you meet the Monkey," he said to me. "You think this place is sick now? Wait." He cleared his throat and did what was obviously supposed to be an imitation of the Monkey, making his voice raspy and tough: "Bobby, do me a favor and answer a phone before I have to stab you."

Besides the Monkey, there was talk of Eddie, another boss, who was evidently out of town, too—or maybe in rehab, it wasn't clear. And someone named Steak Knife, "the big boss."

I didn't know if there was anyone above Steak Knife.

But apparently Steak Knife rarely if ever showed his face, just sat back and counted his money. If there was anyone higher up than him, I wasn't going to be let in on it. At least not yet.

Besides the other bosses, I still hadn't met our reclusive landlord, Morry Krause. The door to his bedroom was always shut, but I could feel the frosty cool of a powerful air conditioner seeping underneath whenever I got up to duck the pink copies of the betting slips in the "trap" at the bottom of the hall utility closet. That was one of the menial tasks assigned me as the new man. It was done so in case we got busted we wouldn't lose all the action we had taken that day.

Of everyone there, Pat was the most paranoid about getting busted. He had a few priors and felt that they might come down hard on him the next time. The office hadn't been raided once in the entire two years it had been going at Krause's place, and everyone felt edgy, the way they do in California when it's been a long time in between earthquakes. There was talk about it being time to move to a new location. A feeling they had been here too long.

Having just started, I felt different. Everything was new to me. I was taking it all in, getting used to things as they were. I wondered about the mysterious Krause and his relationship to these men who came into his home every day to work. Pat spoke of him with contempt. Michael called him "repulsive but intriguing." They showed me a movie script he had written. There were ten green bound copies of it next to the dusty TV. I flipped a copy open and started skimming through. The main character was an old East Village hippie who lived next door to a gang of bikers. The dialogue reminded me of *The Fan Man*, but the plot was even more incomprehensible.

On Krause's bookshelves were hardcover copies of *The Bell Jar*, *One Hundred Years of Solitude*, and *Catch-22*. They looked like original editions, and when I blew the dust off of one, I saw that it was in fact a first edition. It was apparent that these were not the books of a collector but had been bought new at the time of publication. "I wonder if he realizes these are worth money," I said to Michael.

"I doubt it," Michael said. "How much are they worth?"

"I know this Marquez would go for a few hundred dollars, at least."

Michael whistled. "We should just sell them. He'd never notice."

7

A T dinner with a couple of old and close friends, David and Ezra, I found myself telling about my ride up to Winnie's with Michael and what had happened in the couple of weeks since, about the funky office in the East Village and my new job. We were sitting outdoors at an Italian restaurant on Thompson Street. The night was lush, a steamy tropical moistness in the air. Looking at the stylish SoHo crowds in their Agnes B.'s and X-Girls and Marc Jacobs it was possible to think that everyone in the city had money and taste and time to shop.

I suppose I expected my friends to be shocked by my confession. Perhaps I even wanted to shock them. But I'd done too much weird shit over the years—from repossessing cars when I lived in L.A. to driving an ambulance in New York—for them to find it very strange. Ezra said, "It's not like you're telling us you got a job at Goldman Sachs."

"The funny thing is," I said, "in some ways it's not that different." I described the rapid changes in the betting line, the phone madness, the frenzied air of greed. Ezra, who was a journalist, and David, who was a documentary filmmaker, got me to elaborate, to explain the 11-10 commissions, laying off, hedging, scalping.

"Yeah, it sounds like the exact same shit," Ezra said. "Even down to the club atmosphere, the specialized language, the hazing."

The two of them asked many of the same questions I'd asked Michael. How much money was involved? Who was behind it? Was the mob getting a cut?

When I said I didn't know about the mob, they looked at each other. "Don't you think with that kind of money they'd have to be getting something?"

"Michael swears they're not."

"You believe him?"

I pushed up my glasses and fanned open my hand.

"Are you planning to write about this?" David asked.

"Don't ask me. I'm the last guy to ask. Besides, if I write about it they'll probably have me killed."

"What if you get arrested?"

"If I get arrested . . . well, I guess that'll be the end of it."

"They don't prosecute?"

"No. You spend the night in jail, I think, but that's it."

David ran his hand through his hair. Ezra laughed.

"It's weird," I said. "It's not like when I was twenty and soaking up experience. Now I'm just doing this thing, trying to pay my bills, catch my breath—figure out where my life is heading."

"It's what we've come to expect from you," David said.

I laughed along with them, but there was an uncomfortable knot in the pit of my stomach. "Well, I certainly don't want to disappoint anyone," I said.

8

ON my next shift, Krause finally emerged. He came out of the bedroom in his undershirt and a pair of Jockeys that looked like they'd
been on him for weeks. He was wearing a brand-new L.L. Bean jacket.
He had a bad case of bed-head—maybe he really had been asleep in
there the whole time—and his eyes were squirrelly. But underneath
the indoor pallor, graying stubble and sleep circles, you could see he
had once been rather good-looking, maybe even pretty.

Michael introduced us, and Krause studied me a moment. "You
look too cultured to be here with this riffraff," he said. "Surely you
have other options available to you."

I laughed at his Edwardian elocution. "What makes you think so?"

"I'm a very good judge of character," he said. "You have an intelligent
face." He turned to Michael and Bob. "Not like these vermin. How
are the young gangsters? Rubbed anyone out yet?"

"Rubbed anyone out yet," Bob said. "Ha-ha. That's very funny,
Krause."

"You see what I mean?" Krause said, addressing me again.

"Hey, Krause," Michael said. "Did you know that copy of that book
you have over there is worth five hundred to a thousand dollars, according to Peter?"

"Is that right? Which one?" Krause said.

"That first edition of *One Hundred Years of Solitude*," I said. "But it's
more like three to five hundred."

"Those were all my mother's books," Krause said.

"Krause lived here with his mother," Bob confided.

"Don't you dare make fun of my mother."

"A little touchy today, aren't we, Krause? Maybe you didn't get enough sleep."

"Fuck all of you," Krause said. "Except for my friend Peter here." He pulled the blue T-shirt over his belly. "Maybe you'd like to catalog the books sometime. I'll give you ten percent if you can sell them for me."

Everything here seemed to operate on the ten percent system. "I'll think about it," I said.

Krause continued to hover. "How do you like this jacket?" It came down to his bare thighs. "It's nice, isn't it? I just got it by mail order. I really like the pattern."

"You going out in it like that?" I asked, feeling compelled to distance myself from him. I felt that he was being too friendly with me and, given my sense of his standing among the others, it made me uncomfortable. Despite my sarcasm, he continued to talk to me. A circumstance made even more unpleasant by the fact that from three feet away his breath was nearly lethal.

Just at the point where I thought I might gag, he retreated to the bedroom.

Bob fanned the air. "Krause seems to have taken a shine to you."

"Guess I just have an intelligent face."

"Don't worry, he says that to everyone the first time."

After the session, Michael and I grabbed lunch at the Odessa, a Ukrainian greasy spoon on Avenue A. I didn't feel quite comfortable with our roles as employer and employee, but I tried to skirt that issue by playing the reporter.

"So what *really* happens when someone doesn't pay?" I asked, stirring some sugar into my coffee. We were in a back booth, out of earshot of other customers. I'd asked Michael about this once before and he'd told me that if someone stiffed them, they wrote it off as a business expense. But looking at the weekly figures on the sheet Spanky'd given me, and seeing some individual losses in excess of thirty thousand dollars, I needed more convincing. "I mean, that's a lot of dough to write off as a business expense," I said.

"Well, for one thing, collecting money is the agent's responsibility," Michael said.

"All right, but you and Bob have got your own sheet. The Topsider sheet. What happens when one of *your* players doesn't pay?"

"I'm telling you, we write it off. Really. We've had to do that a number of times. The last thing we want to do is make trouble or antagonize somebody. They've got our phone number. If they want to drop a dime on us they can cause us a lot bigger headache than we can cause them."

"Thirty thousand bucks is a lot of money, Michael."

"I'm not saying we don't try to get it. But breaking kneecaps? That just isn't done anymore."

I still wasn't convinced. "So you don't do anything?"

"You think I'm lying?" He stared at me.

I broke the tension by smiling, saying, "No, of course not."

Satisfied, he flagged a waitress for the check, and when it came he picked it up.

9

O NE of my friends from Chicago called to warn me that he'd given
Anna my new phone number. After weeks in the dark, it had
apparently become imperative to her that she have my new number "in
case of an emergency."

"In case of an emergency?" I said. "What the fuck does that mean?"

"Maybe I shouldn't have given it to her," he said.

"What kind of emergency? Her new boyfriend dumping her? That
kind of emergency?"

"I probably should have asked you."

"I mean, that's great," I said, as the idea really sank in. "Now I'll
have to keep wondering all the time if she's going to call me or not."

"I'm sorry," he said. "I fucked up."

"No, no, it's not your fault." And despite my irritation, I was pleased:
apparently I wasn't the only one having a tough time moving on.

Still, after a couple of days of brooding about it, I found that even
that small satisfaction faded. I found myself holding imaginary conversations with her while walking down the street or standing in subway
stations. *In case of emergency!* Christ, Anna. Either let go or don't. But
don't bullshit around with me.

When I picked up the phone one afternoon four days later and heard
a tentative "hi" on the other end, the only real surprise was how much
of an effort I had to make to hold on to my anger.

"Do you know who this is?" she asked, so deadpan I almost thought
she was serious.

"No. I've wiped you out of my memory banks."

"I got your number from Benjamin."

"He told me," I said, only half successful in keeping my voice tight and clipped.

"I didn't like not knowing how to reach you."

"Well, now you know."

"You're furious."

"I'm not furious."

"Pete . . ."

"I'm not furious."

"It's okay. I understand. I've been upset, too."

"It's a little different."

She caught her breath. "Look, I know I said I thought we shouldn't talk for a while. But the last time we talked I felt so pressured. You were talking about moving back to Chicago, and it made me really nervous. Then when you didn't call me on my birthday ... I started to feel like I had fucked up. Like I had made this huge mistake. I really thought you would call me on my birthday . . ."

"Well, happy birthday," I said.

"Pete, really. You don't know how bummed out I got when you didn't call. I just went into this funk."

"You weren't with your new boyfriend?"

She laughed her wicked laugh. "Well .. ."

"He must have enjoyed that."

"He kept saying, 'I thought you *told* this guy not to call you.' He wasn't very understanding."

"What kind of prick would be that dense, not to understand his girlfriend bumming because her ex didn't call her on her birthday?"

She laughed. "Okay, I was awful. But I don't care. I couldn't stop thinking about you."

Despite my efforts, I could feel my guard dropping. "I wanted to call you, you know. It wasn't like I didn't want to."

"You did? You did think about it?"

"Of course I thought about it."

"I guess I'd made it kind of tough," she said, lowering her voice to a whisper.

"You have a way of doing that."

"Maybe you should just come here and spank me," she said.

"It's not like that hasn't crossed my mind either."

"Mmmmm …" A long, dreamy sound.

"What?"

She sighed. "You sound so familiar to me."

My turn to sigh.

"It's hard not talking to you," she said. "I keep thinking it will get easier, but it doesn't."

"No."

"Can't we just settle this shit, sweetie?"

"As in make up our minds?"

"Yeah."

"What do you think I was trying to do?" I thought of all the conversations Anna and I'd had that were just like this one and how each time it was as if we were addressing for the first time some minor misunderstanding that, once we got beyond it, would have us scratching our heads, like, what was that all about?

"I freaked a little, didn't I?" she said. "When you started talking about moving back... I just hate that it's always so all-or-nothing with us. Couldn't you get that job at *Chicago* magazine? Is that still a possibility? Maybe then I wouldn't feel like it was just me you were coming for. I wouldn't get so nervous."

"I was thinking maybe you could find something here."

"In New York? Oh, Pete. It's not like I'm that crazy about Chicago, but—"

"Never mind. Forget it."

"Don't get mad at me. Are you mad at me?"

"No. Why would I be mad at you?"

"Don't you understand? It's just that after all these years of being unsettled, I've suddenly got a gang here, friends—it's hard to think about moving."

"And boyfriends, too. Don't forget about them." I hated myself for talking that way.

"You're not being fair," she said. "You kept saying yourself that one of us might meet someone. And you know I don't like being alone. I've told you that. Besides, Samuel isn't—it isn't like it is with you and me."

"How is it, then?"

"I've told you. It's not nothing, but it's not a big thing, either. I mean, I'm not going to marry him."

"That's not how you sounded a month ago."

She sighed. "I was scared. You were scaring me. I was confused."

"What if I don't get the job in Chicago? What if I can only get a job here?"

"I don't know. Why? Do you have an offer?"

"I'm just asking."

"If you're offered a good job, I guess you better take it."

"I mean, don't not take it because of me."

"Fine."

"Do you have an offer?"

"More than that." I told her about Michael and the office; it pleased me to shock her.

"Oh, Pete," she said.

"Crazy, huh?"

"God, baby. Even for you."

"Michael supposedly made over a hundred thousand dollars last year."

"Jesus."

"It sure would be nice to have money after all this time."

She was silent for a bit, thinking, as I was, about what it could mean. "If you make that kind of money, you have to promise to buy me a house in the country," she said.

"Somewhere in Maine?"

"On a cliff overlooking the water."

"Is that what it'll take, Anna?"

"It sounds nice, doesn't it?"

"It does."

"Oh, Pete, I can't get into talking this way with you. It's too dangerous."

"No, of course not."

I sulked for a moment, but somehow she defused me, gabbing about this and that, changing the subject until I felt giddy and slightly deranged. The call was vintage Anna: inviting me in and pushing me away all in the same conversation—hell, in the same breath. Even as I tried to figure out why—why she did it, why I let her—I felt my heart pumping, renewed and hopeful. It was out of such slim encouragement that my world was made.

10

A T the next session, the Monkey showed up. He was much shorter than his voice over the phone had led me to expect. Even so, the energy of the room changed when he entered; he seemed to displace molecules the way a much larger man would have. The others greeted him with a backslap in their voice.

Closing the door behind him, he took off his suit jacket and undid his tie. He wore lots of gold—a bracelet, rings, a Piaget watch —and smoky-lensed gradated designer eyeglasses, which softened his broken-nosed gangstery face. When his gaze fell on me, he said, "They teachin' you anything?" Without waiting for a response, he turned and carefully laid his jacket atop the sagging blue couch.

It wasn't until he unbuttoned his shirt and pulled it out of his slacks that I noticed the immensity of the Monkey's tanned Dewar's-and-water belly. It was a classic gut, not flabby the way fat is, but rock-hard and shiny, like he had swallowed something very large whole. Standing in the gloom beyond the fluorescent-lit table, he scratched an armpit and said to Bob in his gravelly rumble of a voice, "What's the story here? Why're you half naked?" Bob was sitting with his pants off in the seat next to Spanky. He had started taking them off because of the hot weather, but now it was a superstitious thing he did every morning because the office was on a winning roll.

"You come in here and rip off your shirt and you want to know what I'm doing with no pants on?" Bob's raspy delivery was a fair imitation of the Monkey's.

"Shirt and pants is two different things," Monkey said.

"How was Vegas, boss?" Spanky interjected.

"Like the inside of a big wet pussy."

Everyone laughed, and Spanky repeated the phrase gleefully.

When the Monkey sat down next to Bob, Bob took a ruler and held it to the side of his stomach. "Jesus, it's almost as big as my dick," he cracked.

Monkey looked at me, taking off his glasses. He had cagey brown eyes with a spark of mischief in them. "You haven't said a word since I come in. That's the way everybody in this office should be. Quiet and polite."

"Hey, Monkey, things were a lot quieter before you came in," Bob said.

"Shaddup."

"Tell him about all the guys you've whacked," Bob said. "How many is it? Fourteen?"

"Seventeen," Monkey said.

I looked from one to the other. Who knew if they were joking or not? The Monkey had the pragmatic affect of a killer. An "it needed to be done, so I did it" sangfroid.

"Imlick," he said, addressing Spanky, holding a cigarette between his thumb and forefinger. "Did you pay those bills yet?"

"I took care of 'em, Pop."

"Are you sure?"

"I took care of 'em."

"You know what'll happen to you if you're lying?" He stubbed out his cigarette. His fingernails were large and pearly, perfectly manicured half-moons. "Imlick? I'm talking to you." Monkey started hacking, an ugly, harsh, nasty cough. He brought up some sputum and hawked it into a brown paper bag, which he crumpled and threw across the room, missing the large plastic garbage can by a good yard.

Spanky watched. "Hey, Pop," he said with genuine concern, "are you all right?"

"Yeah, I'm fine," Monkey said, tapping out another Marlboro. "My aim's off a little. That's all."

"How can you smoke when you're coughing like that?"

"It's easy," Monkey said. "I just light it up and I smoke it." While Bob was formulating the opening lines, the Monkey started taking care of other business, calling people about money they owed. He did not sound very friendly with these people. He did not sound like a man I would want to get a phone call from, whether or not it was true he had killed people. "What do you mean he left the country?" Monkey said. "What the fuck are you saying? He left the country and he didn't tell me?"

Hanging up, he did an imitation of the message-taker quaking at the other end. It was the shivery way Bogie looked in *African Queen* when he had the leeches all over him. Ten minutes later a call came in from France. It was the guy Monkey'd been trying to reach. Bob handed him the phone, and Monkey smiled a little smile and nodded while the guy placated him at international rates. After a while he covered the receiver and said, "Anybody need box seats to the Yankee game tonight?" He uncovered the receiver and said back into it: "You got four, right?"

He made another call, this one having to do with some sort of installation date. In the middle of it he looked up, putting his hand over the mouthpiece. "My wife decides we need new carpeting. So she goes out and buys the most expensive carpet they have at Carpet World. It's this disease she has. Well, I go in the next day and tell them, 'Same color but cheaper.' I know she'll never notice. Not a chance she'll notice. She's a sick person, my wife. The things she wants. She's very sick. I mean, last night she actually wanted to fuck me. Can you imagine?"

The phone rang. I picked it up. I had the sense that Monkey was watching me, not just to see how I was handling myself with a customer, but more analytically, trying to size me up. It must have been tough for him. We were like aliens, two beings from different galaxies who had somehow ended up in the same small room.

After the shift, I walked down the five flights with Bob, whose curiosity was more direct: "How come you don't have a real job, a guy like you?"

"It's complicated."

"You looking?"

I wasn't sure if I should tell him that I was. "Ahh, I'm sending out my resume. You know." The absurdity of my words was punctuated by the clatter of our descent.

"You're a friend of Michael's mom, right?"

"Yeah."

"And you're a writer, too, right? You do some writing?"

I grunted.

"You planning to write about this?"

I hadn't been expecting to be asked quite so bluntly, and I was a bit taken aback. But Bob was offhand, cool, like the idea of it didn't bother him.

"Nah," I said. "You know, I'm just trying to make a little money until something else comes along."

He nodded knowingly. We had reached the bottom of the stairs. He adjusted the strap of the bulging extra-large gym bag that hung from his massive shoulder. "Just make sure," he said, and I braced myself for a threat, "just make sure that when you write about this, there's a smart, funny, handsome guy named Bob in it."

I laughed despite myself. That night Michael called to tell me that starting the next week I would become a regular. No more training pay. I'd get five shifts and regular shift fees. Fifty-five per. He'd made up a schedule with me on it. I was in.

11

O n my first day as a full-fledged employee at a bookmaking operation, walking along St. Marks on my way to the office, I was overcome by panic; the street, the storefronts and buildings, the people walking by or gathered on stoops—all looked strange to me, distorted. I had walked this block dozens of times in my life, but now it seemed utterly foreign. Fighting off a wave of dizziness and a palpitating tightness in my chest, I waited for the panic to subside. It did, but it left me aware in a way I hadn't been before of how close I was to losing my grip.

When I arrived at work, Michael let me in without saying hello. He just opened the door, did an about-face, and walked away. Spanky was the only other one there. He didn't acknowledge me either. The silence was thick and unfriendly. Suddenly Michael snapped at Spanky, "Don't you even care whether the office won or lost last night? Jesus!"

Michael's anger shocked me. This was a side of him I had not seen before. The subtext, which I got from him later, was that Spanky had recently asked him and Bob for a piece of their piece. Spanky'd been working for them a year and felt that he'd earned it.

"We were curious to see what he'd ask for," Michael told me. "We were prepared to give him two percent. He asked for eight. It proved to me he didn't have a clue to what was going on. So we said fuck it and didn't give him anything."

I was tempted to ask why, in that case, Spanky should give a shit about whether the office won or lost. But I thought better of it.

To my unschooled eye, Spanky seemed like a good clerk. Not only did he handle the phone well but he brought other skills to the job: he was a carpenter on the side and had built the worktable and installed the fluorescent overhead fixture and the trap in the utility closet.

Michael knew Spanky from high school and had hired him initially to stand lookout for the cops. They didn't pay him shit, but Spanky was desperate because it was winter and carpentry got slow in the cold weather.

Eventually he started filling in shifts. Now he'd been working the phones for a year. Out of everyone there, Spanky took the most time with me trying to explain things. "You're doing fine," he kept reassuring me. "Just be a little more confident on the phone. Like you're in control. Don't rush things. Make them go at your pace."

It would be a few weeks before his attitude toward me began to change, before I got a glimpse of what it was about him that made Michael so testy. But in the beginning, Spanky was my ally and teacher. And it was Michael who struck me as the prick.

12

L ESS than halfway through my first official week, the editor I'd met from *Esquire* called me back. He'd read my query letter and was intrigued. Would I like to have lunch with him and tell him a bit more? I immediately started to get nervous. Monkey claimed to have killed people, for Christ's sake! What if he turned out to be not so keen on reading about himself in *Esquire?* I agreed to the lunch anyway. I needed desperately to think of myself some other way than as a down-at-the-heels thirty-three-year-old underachiever who was breaking the law to pay the rent. I'd decide how to play it when I was sitting across from the guy.

Of all the characters at the office, Bob was the least reticent around me. The others regarded me with what might charitably be called suspicion. Bob did, too, but in a humorous way.

The day after the call from *Esquire*, I went in to work and it turned out to be just the two of us. There was only one day game, which he could have worked himself, but, as he said, I needed the experience.

"Let me tell you something about this business," he said after I locked the door behind me. "Don't believe anything anyone tells you."

"What do you mean?"

"Everyone's got some deep, dark secret that they're hiding."

"Yeah?"

"It's hard to explain. But stick around, you'll understand eventually."

"So what are you hiding?"

"Me? Nothing." He laughed. "See what I mean?"

I laughed. "No."

He jabbed out a phone number, and while he was waiting for an answer, he said, "You know anything about playing the guitar?"

"Not a thing."

"Me either. But I think I'm going to learn."

"You play any other instruments?"

He ignored me and spoke into the phone: "This is a message for Grant. My name is Bob. I got your name from the *Voice* about guitar lessons. Please call me at the this number. . . ."

After hanging up, he said, "I set goals for myself. Each time I reach a new plateau financially, I set a goal for myself. It's good. It keeps me growing. I've never played any musical instrument and I think I might be tone-deaf. But what the fuck, you know?"

"Hey, I think that's great."

"I don't know if it's great. But I figure, what good is money if you don't spend it on something?"

"Have you ever thought about another line of work?" I asked. "Besides this?" I motioned around Krause's dreary quarters.

"Me? Nah. I'm a business guy. This is what I'm good at."

"But why bookmaking?"

"Because I'm not smart enough to go into real business."

"C'mon, I'll bet you could do anything you want," I said.

"I suppose so. But if I was working as a trader, you think I could go to the office like this?"

It was true. His shirt was open; he wasn't even wearing pants. Stripped down to his lucky boxers again.

"I wouldn't mind having my own trading operation," he said. "You know, where I could just work on a computer out of my home. But going into some office in a suit? I don't think so." Closing up afterward, he told me that the night before we had done $120,000 worth of business. "Not bad for baseball, but not like football season."

I was thinking about all that cash. "How do you pay people off and collect?" I asked.

"What do you mean?"

"I mean physically. Do you actually meet somewhere with a paper bag full of money?"

"Sometimes. Sometimes we do electronic transfers. Mostly it's done in person. Why?"

"So you mean you've gone out, walking down the street with thirty thousand dollars in a brown paper bag?"

"Sometimes more."

Bob was six feet one and built like a linebacker; he boxed three times a week at the Times Square Gym. He wasn't exactly the stuff of a mugger's dreams, but still, that was a lot of dough to carry down a New York City block.

"Don't you worry about getting ripped off?"

"Nah."

"You ever carry a gun?"

He grinned. "No. But Pat does. Or at least he says he does. Anything else bothering you?"

"Actually. .." There was something else I had been thinking about: why bookies charged customers a ten percent commission on losing bets but nothing on winning bets. It didn't make any sense to me. Why didn't they just charge five percent across the board?

"You ask a lot of damn questions," Bob said. "You sure you're not a cop?" It was the second time he'd asked me that in the past couple of days, and though I was pretty sure he was joking, the strange thing was that I couldn't look him in the eye when I tried to brush it off.

All those stories about guys who bluff their way through a lie-detector test—I was the polar opposite. I could be asked, "Is your name Peter?" and my "Yes" would make the needle jump off the chart. My natural state was one of extreme guilt. I lived in terror of all the things of which I might be guilty. So I'd taken to confessing; I was a compulsive confessor. I told people about things I had done before they got a chance to ask me.

I wanted to tell Bob about my *Esquire* lunch. I knew I shouldn't, but I wanted to. And my reasons went beyond guilt. They went beyond being forgiven blindly for all my sins. I wanted him to know that I wasn't just some schnook. I was somebody. I had something going.

Fortunately he looked at his watch and said, "Hey, Pete, it's been swell. Gotta go." And my big mouth was left hanging open.

13

A ROUND this time one of my best friends, Charlie Rosenbloom, stopped off in New York on his way back to L.A. from an assignment in England. I met him after my shift, and we ate a late dinner at a Vietnamese place on Bayard Street. Afterward we wandered into Little Italy for espresso and dessert at one of the outdoor cafes on Mott Street. Like me, Charlie was approaching his mid-thirties uneasily, having failed to live up to his own and others' expectations. His marriage was rocky; his work less and less satisfying. Whenever I'd seen him over the years I'd always been able to gauge his general mood and outlook by the length of his hair. Short meant good. Long meant depressed. Facial hair meant manic and crazy. On this visit his thick black hair was on the long side, though not quite ponytail length. Basically, Charlie was struggling to get by and feeling a bit battered and world-weary.

It was good to have someone to commiserate with, someone who didn't require elaborate explanations, someone who understood. It made life seem less inexplicable, less lonely.

As often happened when the two of us got together and were grasping for ways out of our "situations"—or at least ways to improve them —we discussed collaborating on a screenplay.

In the past our off-the-cuff movie ideas had been good enough or wacky enough to get us giddy over our beers even if deep down we knew we were merely amusing ourselves with get-rich-quick schemes on which we'd never follow through.

Our talk now had a different tenor. We were both feeling desperate. Time was going by. Charlie was on the freelance treadmill in L.A., and

I was . . . well, you know about me. When he told me his idea for a wrong-man suspense-thriller in the Hitchcock mold, I think there was a feeling that together maybe we'd be able to prop each other up and overcome self-doubt in a way that we weren't able to do on our own. I latched right on, got excited, tossed off a couple of possible twists to the story.

We paid the check and walked the streets, talking still, our animated voices carrying us past fish markets with large half-dead carp swimming in plastic buckets and store window displays of glazed ducks and shelf upon shelf of fried yellow pastries. Somewhere near but out of sight firecrackers popped. Charlie told me if I was serious about it, he was coming back through New York in a couple of weeks for some rounds with editors, but he'd stay on so we could get together and hammer out a treatment.

Suddenly lighter of heart than I'd been in months, I said, "You mean it?"

And he said, "I really think we can sell this."

"It's definitely salable," I agreed.

"So the only thing standing in our way is us."

The lightness had passed quickly. I felt something else stirring. Tears gathered in my eyes. Charlie looked at me, and I shook my head. "Man, Charlie, what is it with us?"

"Hey, this stuff isn't easy."

"But we make it so hard."

"Yeah, we do," he said. "We sure do."

"Why?" I asked. "Why do you think we do that?"

"I have no idea. I just know we've gotta stop."

14

I was the first one to arrive at work the next morning, able to let myself in with the set of keys that I had finally been awarded. Krause's apartment was dim and cheerless, the blinds drawn as usual. I plugged in the light and watched it flicker on, revealing the previous night's storm of business: ashtrays piled high with cigarette butts, baseball "books," which listed the official Las Vegas rotation—or order —of the games, broken pencils, date-marked cassette tapes, rubber bands, pink betting slips, and the off-the-hook handsets of all six phones. I straightened up, using a newspaper to sweep ashes off the polyurethaned tabletop, trying not to breathe in too deeply the sour smell of cigarette smoke that lingered in the air.

Pat came in, grumbling under his breath, while I was in the midst of my tidyings. He sat down and opened his *Daily News*, pointedly ignoring me. When Bob and Michael came in ten minutes later, Pat said, "I got carried out yesterday. *Carried out.* A fuckin' disaster."

Bob said, "So what else is new, Paddy?"

"I'm not lying. I got *carried* out. Next time you see me I'll be getting on a plane to Argentina."

Bob turned his attention my way. He told me he'd been up all night thinking about my question as to why bookies didn't just take a five percent vig off of winners and losers instead of ten percent off losers. He said that he thought it wouldn't work as well because there were more losers than winners. He scribbled some figures on a piece of paper to prove his thesis.

Michael looked at the paper and said, "That's ridiculous. There's going to be just as much money on one side as the other." He and Bob had a big argument over it while Pat and I looked on in amusement.

Pat said with acid matter-of-factness, "There's only one system that's ever worked and that's eleven-to-ten. Bookies have been using it forever, and there's a very good reason: it works." Michael and Bob continued to argue until at last Michael looked at Bob's figures and his expression started to change. "Look, look! It's beginning to dawn on him. It's beginning to penetrate the skull matter. Do you get it now, Einstein?" Bob gave me a high five.

Michael said, "Of course there's going to be more on one side than the other. You have more losers than winners."

"Exactly," Bob said. "That's what I've been telling you all along. There's more losers than winners. It's just human nature. The guys who stop playing, they all stop because they're losing. You ever see a winner stop? No. It never happens. People only quit as losers. See, we've got all these guys who've quit or we've cut 'em off, and to a man they're losers. It's dead money in the pot. But the guy who's winning? He doesn't stop playing until he's a loser, too. That's just the way it works. That's the real reason we only tax losing bets. There's more of them. By taxing them we make more money."

"Eleven-to-ten," Pat said. "Ask Steak Knife why he's richer than God and he'll tell you: eleven-to-ten."

The mention of the big boss—Steak Knife—whom I still hadn't met, stirred my uneasy and thus far unsatisfied curiosity. I continued to find it hard to believe the Mafia wasn't getting a piece of our profits.

I asked Bob about it, as I had asked Michael, and like his partner he denied the connection. "They're not into it," he told me. "We don't get involved with any wise guys. Steak Knife's been in the business for thirty years and he's got his own juice. But he's not a wise guy."

15

D o you believe them?" Anna said.
"I don't know." I was in my underwear, lying on the couch, the
receiver crooked in my neck. Out the window, across the dark slick of
the East River, I could see the flickering towers of lower Manhattan
and, farther along, the jeweled lights of the South Street Seaport and
the Brooklyn Bridge.

"You don't know or you don't want to know?"

"Both."

"You sound tired."

"I am. It's not easy being around those guys."

"Macho?"

"Yeah."

"That must get tedious."

"It does, a little. But I like it, too. What happens there is interesting."

"Because of the money?"

"No. Because it's a new world to me. That's all. Hell, the money,
Anna. I'm making nothing at this point. I might not even be making
enough to afford an apartment of my own. I'll have to work a year
there before I start to make anything real. And even then . . . It's not
just putting in time. You have to bring in players, too, if you want to
be a real part of it. It's like a law firm. The associates who get made
partners are the ones who bring in business."

"How do you bring in business?"

"I don't know. You solicit it. You have to hang out in places where
there are gamblers."

"What about your poker players?"

"Well, I haven't been playing much. None of them really bet, anyway. Poker players and gamblers are different. Serious poker isn't about gambling so much as it's about skill."

"So poker players don't bet?"

"Not on sports. Not most of 'em, anyway."

As in our previous conversation, we talked easily, intimately. There was an undercurrent of frustration, at least on my part, that we could talk so well, feel so linked, and not be together. I tried to take it for what it was (whatever it was) and not put any pressure on her for some kind of resolution.

16

I⊤ turned out I wasn't the only one in the office losing sleep over a woman. One night Michael dropped his usual affect of tight-lipped cool to tell me about "this incredible girl" he had just met.

"I thought you had a girl," I said. "What about Lindsay?"

"We broke up."

"Really? That's too bad."

"Well, you're definitely in the minority on that. Right, Bob?"

"No, I liked Lindsay," Bob said, but the moment Michael looked away he recoiled and made a pukey face.

"Anyway," Michael said, "I can't stop thinking about this other one."

"The one you just met?"

"Yeah."

"There's only one little problem," Bob interjected. "She happens to be the fiancee of our best customer."

Bob and Michael made it a practice to take their biggest players out to dinner periodically, usually to Ben Benson's or Luger's or some other wise-guy steak joint where they could wave a lot of dough around and act like big shots. "It's good business," Bob liked to say. "These guys play us some big numbers, so we give a little back. It pays off. A couple of times already we've gotten new players out of it, either guys they brought along or guys they referred to us."

"Do you think this guy considered his fiancee in the nature of a referral?"

"I don't think Tuna would have seen it that way," Bob said. "If he had noticed what was going on."

Michael smirked.

"Tuna," I said. "I've taken bets from that guy."

"You've done more than that," Michael said. "You've met him."

"I have?"

"At that poker game I took you to. He was Tony. The one who looked like a dark-haired Fabio."

"That was Tuna?"

Michael nodded.

"He's a fucking idiot," I said.

"Exactly."

The guy Michael was talking about was one of the worst poker players I had ever seen. After he'd dropped a grand and split, I had asked what kind of drugs he was taking, but Michael had said no, the guy was just incredibly stupid.

"Where does a guy like that get the kind of money he's betting here? Is he a trust-funder?"

"He's a fence."

"You mean a guy who sells stolen goods?"

"Luckily for him, it's work that requires no brains," Bob said.

"So what's this girl doing with him?"

"He's rich and good-looking," Michael said.

"That doesn't say much about her."

"She's a goddess," Michael said. "Don't you dare insult her."

"Yeah, easily worth five thousand a week," Bob deadpanned. It turned out that was what Tuna was losing on average.

"We were sitting there at Sparks," Michael said, "and all of a sudden I felt her hand on my knee."

"She had her hand on your knee, too?" Bob said.

"Don't listen to him," Michael said. "He's just jealous because he's married." Bob had been with his girlfriend, Stacy, for three years.

"Listen to me. You want to make up Tuna's weekly nut, you're welcome to do whatever you want. Just so it's clear to you that it ain't coming out of my pocket."

"You have to admit she'd be worth it."

"We're talking about five grand a week."

"I really think she's the best-looking girl I've ever seen," Michael said.

"Five grand a week."

"I haven't done anything yet," Michael said.

"You will."

17

I met Eric Levitt, the *Esquire* editor, at the Royalton at noon on a Friday. He'd wanted to meet me on the early side because he was heading out to Sag Harbor after lunch.

I was apprehensive and a bit nervous, and the setting didn't help— publishing bigwigs holding court in their royal blue palace. Levitt was a tanned, intense, prematurely balding guy in his thirties with thick black eyebrows that nearly merged and quick black eyes.

In the query I'd sent him, I'd described Michael and Bob— without naming them, of course—and I'd described the level of their education. I hadn't mentioned my own involvement, but I hadn't been thinking about a piece like that. After a couple of glasses of Chardonnay, how-ever, I couldn't resist telling Levitt everything. As I talked, I could see the wheels turning behind his intense gaze.

"Are you still thinking about doing a piece, then?" he asked.

"I'm not sure."

"I could see where you might have some problems."

"Yeah."

We sat there, each of us considering what those problems might comprise.

"Were you thinking in terms of first person?" he asked.

"Yeah. I'm just not sure quite how it would work. I wouldn't want to use a pseudonym."

"It seems to me that you'll want to wait a while anyway. You've only been doing this a few weeks, right?"

"Right."

"There's really no rush. Why don't you just see how it develops and then let me know what your feeling is further down the road?"

That was pretty much what I had been thinking anyway, but I realized I'd also wanted some kind of assignment from him. Something definite. So I could feel like there was a plan in place, a purpose.

"By the way," Levitt said, "what if I wanted to make a bet with you guys? What's the minimum?"

I laughed. "Seriously?"

"Seriously."

"Fifty bucks is the minimum bet. If you're interested, I can give you the office number. What do you bet? Baseball?"

"And football when it starts."

"You really want the number?"

"Please."

I opened my wallet and found the scrap of paper I'd written it on. "You'll need to come up with a name," I said.

"A name?"

I explained to him about the code names.

"I like that," he said. "Code names."

"You can call me when you decide on one," I said.

I will," he said

And that was how my lunch with Eric Levitt the *Esquire* editor ended. Not with an assignment but with me giving him the office phone number.

Great.

18

Yᴏᴜ got a player!" Bob said. "All right! That's how it starts!" He high-fived me with a bit more enthusiasm than I returned. It was just the two of us, working the day session. "Who is this guy? Is he good for his money?"

"Yeah."

"You sure?"

"I'm sure."

"He better be, because you know who has to make it up if he doesn't pay."

"Yeah, I know."

"What name did you tell him to use?"

"I haven't yet. I was thinking of Esquire for Curly."

"Who's Curly? You?"

"Yeah."

"What are you, crazy? You think you're getting your own sheet for one player?"

"I don't know. How does it usually work?"

"The way it works is you don't get a sheet for one player. If you want, you can put him on our sheet, the Topsider sheet, and then we give you ten percent of whatever he loses. But this Curly stuff—that's a ways off."

My enthusiasm was waning.

"What exactly don't you understand?" Bob asked.

"I get ten percent?"

"Right."

"And for that I'm totally responsible for anything he doesn't pay?"

"Right."

"And you think that's fair?"

"That's the way it works."

"When do I get more than ten percent?"

"More?" He laughed. "Get some more players, motherfucker, and we'll discuss it."

"So what do I tell this guy?"

"Tell him his name is Esquire for Topsider."

Bob pulled up the antenna on his pocket phone and called Michael. "Hey, baby. Guess what?" he said. "Pete's got us a player." He gave me the thumbs-up sign. "Yeah, he knows. . . . What? Tonight? All right." He folded up the phone. "Hoo, baby. Are you working tonight?"

"Yeah."

"You've never met Eddie, have you?"

I shook my head.

"But you've heard us talk about him, right?"

"This is the guy who was in rehab?"

"It wasn't really rehab. He just went away for a while to pull himself together."

"So he's coming in?"

"Tonight."

"You don't like him?"

Bob shrugged. "He's one of the partners. I don't know what it is. Maybe because he's kind of a screwup, he tends to have a chip on his shoulder. Like the first time I met him," Bob said. "It was my first day working—I'd gotten in through my uncle—and so there I am, the new guy, and in walks Eddie. He throws his car keys at me and says, 'Go park my car, kid.' But I didn't move. I said, 'I'm here to be a clerk, not a parking valet.' He said, 'Well, in that case, just take a look at the car as you go out the door, because you're fired.' "

"You're kidding. So what'd you do?"

"Nothing. He couldn't fire me. My uncle was friends with Steak Knife. Monkey told Eddie to go park his own car."

I laughed, but when I showed up that evening, there was this lanky long-haired balding guy sitting in the boss's chair, and before I was even introduced to him, Spanky said to me, "At six-fifteen Eddie would like you to go down and put a quarter in the meter for him." Spanky gave me a quarter, and Eddie said in a hoarse voice, "It's the Chevy Blazer with the Colorado plates." I had the feeling that Spanky had been assigned to this chore initially and certainly would have done it if I hadn't been there. Since I didn't have Bob's uncle behind me, I didn't say to Eddie what Bob had. Eddie had a gambler's twitchy nicotine face, with long, thinning blond-gray hair pulled back into a ponytail, and aviator glasses, which I later found out were Cartier's worth over a thousand bucks. There was an amphetamine jerkiness to Eddie's movements. His lips quivered. When Michael asked him to answer a phone at one point, he said, "I don't answer the phones anymore. I did that once. I don't do it anymore."

I went down and put the quarter in his meter at six-fifteen. When I got back, Eddie asked me if I'd taken care of it. I said, "Yeah. The red Camaro with Connecticut plates, right?" He looked startled for a moment until he realized I was playing with him. It didn't win me back my pride, but it got a laugh out of the other guys and it was better than nothing.

Michael was all excited on this particular evening because Bob had cut a deal in the afternoon to take over the bettors of some bookie who'd just retired. The bettors happened to be Chinese. The deal was evidently a coup of sorts; it meant a lot more money. To me it meant something else, too: trouble.

"Aren't the Chinese all tied up by the gangs and the tongs?" I asked.

"No, this guy had been in business for years," Michael said. "It's something established. We're not coming in and stepping on anybody's toes."

"How do you know?"

"I know. There's nothing to worry about."

I wanted more reassurance than that, but Michael changed the subject. He told me he had seen Shelly, Tuna's girl, for a drink.

I said I wasn't surprised.

"You think it's stupid?" he asked.

My relationship to Michael at this point was peculiar. I was nearly ten years his senior, a good friend of his mom's, and yet I was now working for him, making a clerk's wages while he was earning serious money. It had altered the balance, put things off-kilter. Yet there were still levels on which he sought my approval.

"You think it's stupid," he said.

"Didn't you tell me this guy was your best customer?"

"He is."

"So what do you want me to say?"

"If you saw her . . ."

"I know she's beautiful, Michael. They always are."

"What do you mean, 'they'?"

"The ones who get you in trouble."

He laughed. "I'm not worried about trouble. It might just be very expensive."

"Well, then, I hope she's worth it."

19

CHARLIE Rosenbloom called me from the road to see whether I had been giving the movie idea any thought.

"Yeah. Are you coming back through town?"

"Next week. I'll be staying for a while, too. You ready to do some work?"

"Absolutely. The only thing is, we might have to work it around my shifts at the office. I'm a full-fledged employee there now."

"Really? What's the line on the Yankees tonight?"

"Very funny."

"Pedro, I don't see what your problem is. That's a great gig."

"I'm a fucking criminal."

"Don't be so status-conscious."

It was hard. Outside of the office I was already finding it difficult to integrate my two lives.

My close friends were one thing. But acquaintances, people I knew from the literary and magazine worlds, were another story. At the couple of cocktail parties I went to, I found myself squirming when asked what I was doing now. I couldn't look into anyone's eyes. I hadn't resumed writing yet. I was still without an apartment of my own. There was this woman in Chicago and . . . and I was, of course, each day breaking the law.

The people at these parties were mostly in their thirties, their lives in midstream; they talked about work and publishing, summer homes, mortgages, private schools for their kids. It filled me with a sense of shame to listen to them; there was no way to explain myself, my tripping-over-my-own-feet fall from grace. No way to tell them the truth without feeling as if I were some sort of living object lesson. An example of what not to be.

Most nights after work when I didn't go out, which was most nights, I would return to the studio in Brooklyn, make myself a sandwich or a bowl of pasta, and start reading one of the several books on screenwriting I'd purchased. I was always looking for a map, for some kind of guide, but in a way, the impulse was probably just another form of my lifelong habit of procrastination: one feature of maps was that you could read them without actually having to go anywhere. You could pretend you were planning your trip—and you might feel as if you were —but you never even had to get off your couch.

By eleven o'clock I'd be tired and lonely for the sound of some voice other than the one in my head. I'd resist calling Anna by turning on the news, then Letterman. Sometimes I'd nod off before I could muster the energy to make it to bed. It was hard for me to decide if this sort of behavior constituted depression. I didn't *feel* depressed. But that didn't mean I wasn't.

I asked my friend Ezra if I seemed depressed to him.

He said, "No. Not like you need to be hospitalized or anything."

"Well, will you let me know when you think I do?"

"Sure. But why don't you just start taking Prozac like everyone else?"

20

THE payoff from all the resumes I'd been mailing out was exactly one response: the publisher of an alternative newsweekly in Connecticut wanted to see me about the top editor job. It wasn't anything I had ever imagined, but then, at this point what was?

I tried to visualize myself in Fairfield County. Would Anna like that? She had her fantasy about living in the country with a big dog. Would this qualify?

I kind of liked the idea myself. But what if I couldn't persuade her to join me? Then what? Me and the dog alone in the burbs?

And what would happen to the script with Charlie? A job like that with a small staff and budget would be extremely demanding. Where would I find the time?

I called Anna just to get a reading on her. Or maybe because I wanted an excuse to talk to her. It had been a couple of days.

"Aren't you getting a little ahead of yourself, hon?" she asked in a voice that still had a touch of sleep in it.

"Don't I always?"

"It's a sweet idea," she said. But that was about as enthusiastic as she would get.

She was right, after all. Why worry about it? It wasn't as if I had been offered the job yet.

I just knew that I would be.

After I got off the phone with her, I went downstairs and picked up another tenant's blue plastic-wrapped *New York Times* from the floor of the vestibule and took it outside.

Columbia Heights was mercifully cool, the sun not yet having made its way over to my side of the street. I sat there on the sandpapery brown stucco stoop in paisley shorts and a T-shirt, unshaven, a bit woolly-eyed, studying the baseball box scores. After a while I looked up from the paper, and Anna crept back into my thoughts. It wasn't surprising, really. Being out on that stoop put me in mind of the wooden back-stairs porch of the apartment in Chicago where she and I had often sat on sunny mornings, drinking coffee and discussing the day's news.

More and more, I had a tendency to gloss over the problems we'd had, remembering only what I missed. Whatever Anna's difficulties, they didn't obscure the fact that she *did* love me, and I liked being loved by her. She was a wild, willful girl whose emotions and sex drive ran high, a Catholic girl who would bless me with her guilt by saying, "How come I feel so shameless with you?" And she was blessed herself, by beauty: a mischievous, sly, wide mouth, small breasts with huge nipples, a warm, lovely bottom that she liked me to rub, and of course those absolutely clear green eyes that could look right into me and take my breath away. What other woman had I ever known who used terms like *teleology* and *vitalism* when talking about love? Or who would describe men as the state and women as guerrillas in the surrounding jungle? Or compare herself to the French revolutionaries, always trying to start everything at the year one?

But it was more than that. Anna saw and named the things in me that needed naming; she got me. When I thought about love, just love itself, all the other stuff became trivial. Her inexplicable rages, the crazy self-destructiveness—the something inside her that made her want to "fuck up and ruin things." I could summon whole scenes of lunacy—a fight we'd had in her car, driving on Lake Shore, that ended with her suddenly yanking the keys from the ignition and throwing them out the window; the guy she swore she hadn't slept with pounding on the back door of the apartment late one night, drunk and shirtless, demanding to see her; the time she quaffed an entire bottle of Tylenol in front of me and I had to take her to the hospital to have her stomach pumped; the way she'd sometimes call me from work to tell me that she was lying on the floor of her office and needed me to come get her and take her home—I could summon these scenes, but they were like stories I had read somewhere, funny and weird, no longer painful.

A suited man with a briefcase came out of the door from behind me. I looked up. He nodded without smiling, eyeing the sections of the paper that lay beside me. I watched him go down the block in the direction of the subway, and my stomach filled with butterflies. Something about his purposeful stride.

I reassembled the *Times* and returned it to the cool dirty tile floor of the vestibule. Then I went upstairs to change for work.

21

IT took me a while to figure out why some days when I arrived at the office everybody was nasty and glum. Since I was just making shift pay and had no direct interest in the office's fortunes, it didn't occur to me how closely the guys' moods mirrored the previous night's results.

Take Pat. He'd been depressed and grouchy for days now because he was losing. He was muttering to himself all the time like a crazy person, "I'm getting murdered. I'm getting killed." It was the way all gamblers talked about losing—like it was akin to death. "Four to three top of the ninth. My life is over."

Of course a certain amount of self-hatred accompanied us into the office that had nothing to do with wins or losses. At least those of us who were still not able to rationalize what we were doing with our lives. Perhaps that was only me. The fact is, things usually warmed up when we began to work, when the promise of money took hold.

A fair portion of the surliness was just macho posturing, verbal towel-snapping. As in a fraternity, the brotherhood part rang false to me—justification for the meanness of initiation rites and hazing rituals.

During a lull in the action, Eddie sent me out to get a lemonade extra sweet and a chocolate egg cream. I didn't mind at first, but on my way back to the office, carrying the drinks in a brown paper bag —the same way I knew I'd be carrying $30,000 if I stuck with this— I caught a glimpse of myself in a store window and was startled: it wasn't anything in the way I looked, really; it was just the shock of seeing myself reflected. Way to go, kid, I thought. Mom would be proud.

When I handed Eddie his order, he peeked inside the bag and said, "Where are the straws?"

"They're not in there?"

"Good job, Pete," Spanky said.

There were snorts of laughter.

"Didn't they teach you anything at Harvard?" Pat said.

I bit my lip, murmuring something under my breath about being a "fucking thirty-three-year-old delivery boy."

Spanky must have heard me. He said, "Hey, we all had to do it, Pete. If this is the worst thing that ever happens to you in your life, that you have to go out and get some sodas, you should consider yourself lucky."

A while later, looking at Pat, who was still muttering about his mounting losses, his face flushed and sweaty, I thought about luck. I thought, What would Pat feel like if he lost his health? If he were faced with real death and not just its gambling equivalent. Would he think that he had wasted his life?

Or was that precisely the question that all the little metaphorical deaths of sports betting allowed him to avoid?

And what would my answer to that question be?

22

WHEN the Monkey heard I was looking for an apartment, he proposed an arrangement: "Get yourself a place downtown. We need a backup office in case we run into any problems. I'll pay you a hundred dollars a week, plus take care of your phone bills. Then, if we need to use it, you'll get two hundred a week."

I said no as quickly as I could. But Monkey worked on me. He said, "You'll get a nice place. What's the big deal? This is just till you get back on your feet."

I said no again. He shrugged and let it go.

Afterward I started thinking about his proposition. I was looking through the classifieds, and it was impossible not to fantasize. With the extra money, I could get a place in the Village. Subtract four hundred dollars a month and I was way out in Brooklyn somewhere.

I started wondering what it would mean for me if we got busted and my name was on the lease. I needed to find out. A lawyer acquaintance said, "Legally, it wouldn't make things worse. But if you're asking for advice . .."

I was getting plenty of that. One of my poker buddies, Melman, who worked on Wall Street, decided to tell me a cautionary tale about a guy he knew from another poker game—"nice guy, sweet guy"—who got involved in something to do with illegal slot machines and got arrested. Now he had a trial upcoming and a jail term facing him. "He's scared shitless," Melman said. "He's so fucking scared, Peter. You don't want to know about fear like that. You really don't."

Anna also cautioned me. She said, "Baby, get your own place. It just isn't worth it to get that tangled up with them."

I knew they were both right. But I couldn't let it go. I called a couple of real estate agents and arranged to see some one-bedrooms in the Village. I just wanted to see what was out there.

23

I met Charlie at the place he was house-sitting in SoHo. Somehow he always managed to land great digs for his New York visits. This one belonged to a Chilean artist couple he knew, and their paintings, which Charlie told me were collaborations—a good omen—filled the apartment like big bouquets from Latin America. We set up shop in the dining area, a long, narrow space with a big table that gave us room to spread out.

Charlie had a yellow legal pad on which he'd already made some notes, and another one that was blank. I'd brought along my own notes. There was an air of seriousness to the proceedings.

To get us going, Charlie talked the story through as far as he'd gotten. I listened, then made observations. I asked a lot of "what if?" questions: "What if the killer has some identifying mark, and when the hero identifies this other guy, the wrong guy, it's because he has the same mark?" Soon we moved from the business of plot to character. To know what was going to happen to these characters we had to know more about who they were. "What does this guy do for a living? What does he want? What do we know about him?"

It was hard work. We had our basic concept, but now we had to flesh it out, make it real. And getting one level deeper only made us aware of how much further we'd have to go before we knew anything. Really knew anything.

Charlie wanted to stay away from the autobiographical. "Whatever we make the hero, he isn't going to be a writer. That's definite," he said. "I hate it when people make a character a writer. It's just laziness."

I'd always been afraid I lacked imagination. That I had trouble fully conceiving a life different from my own. It was part of the reason I'd thought that Charlie and I would work well together. But he was so dogmatic it unnerved me.

After a few hours we decided we'd done enough for our first night and went out to Nick and Eddie's for a drink.

"This autobiographical thing," I said.

"What about it?"

"It's going to be there no matter what we make this guy."

"That's not what I was saying."

"You're just saying we shouldn't make him a writer."

"Right."

"Why don't we make him a bookie?" I said with a smile.

"That could be interesting."

"I'm kidding."

"Why not make him a bookie? I like that."

"I don't want to." I was surprised by my vehemence.

"Why not?"

"I don't know, I just don't."

Charlie looked at me strangely but left it alone. Later on, it would occur to me that I felt protective of my life at the office. It was mine, and I didn't want to share it with him.

"The main thing is," I said, "that this guy, our hero, whatever it is he's doing, isn't a fully realized person, and then this event happens, this galvanizing event, and—I mean, isn't that what both of us have really been waiting for ourselves?"

"But we're *trying* to make something happen."

"Yeah, I know, but. . ."

A couple of young SoHo ingenues came in and took seats down at the other end of the bar.

We both looked.

"But we still hope it just happens," I said. "I mean, I guess I do. Something just falls out of the sky."

"When was the last time you got laid?" Charlie asked.

"A long time."

"How long?"

"Not since Chicago."

"That was nearly a year ago."

"I had a couple of chances last winter. I even got into bed with this one woman."

"And?"

"I don't know. I couldn't. Her bedroom smelled like air freshener."

"Air freshener?"

"She had these Glade plug-ins. You know, these things you plug into the wall?"

"Yeah?"

"And somehow, well ... I just couldn't. ... I kept thinking about Anna."

Charlie put his hand to his head. "When was the last time you talked to Anna?" he asked.

"Uh ..." I did my best to swallow the word. "Yesterday?"

"No!"

"We've started talking again."

"I thought she told you she wanted to move on."

"She thinks she might have made a mistake."

Charlie's disgust rendered his whole face slack. "You better be making enough money off this bookie stuff to afford a good therapist."

I laughed.

"No, I mean it. You are seriously fucked up."

"Stop."

"I hope you're not considering getting back with her."

"It's not really possible at this point. Geographically."

"You better promise me that you aren't going to see her while we're working on this script."

I didn't say anything.

"Pedro, seriously, dude, what are you doing?"

I flexed my hand, looking away from him.

"How many times are you gonna fuck yourself up over the same woman?"

"I don't know how to explain it to you," I said.

"How do you explain it to yourself?"

I opened my mouth. But no words came out.

24

Monkey arrived at the office to a chorus of "Hey, boss" and "Hiya, Pop." Me? I just nodded. I couldn't quite get around to calling anyone "boss." It made me feel too much like one of the bad guy's flunkies in an old *Superman* episode. The odd thing was that the few times Monkey had rung up the office from outside and I had answered, he had called *me* "boss." He used the word the way some people used "pal."

At all the commotion, Krause poked his head out from his bedroom, squinting, half asleep. "Can't you cretins show a little more consideration? Do you have to yell and scrape your chairs around?"

"Whadda *you* want?" Monkey growled at Krause. "Go back into your cave."

"We've been getting complaints from the guy downstairs."

"So? Throw him a few bucks."

"That's how you guys solve everything, isn't it?"

"And when was the last time you didn't take the money?"

Krause recoiled, momentarily wounded. "Fine. He can call the cops. I don't care." He slammed his door.

Monkey stripped off his shirt. He wore a ribbed tank top underneath. "It's too hot in here," he said. "That air conditioner is no fucking good. We're going to need something bigger."

"A new filter will take care of it," Spanky said.

"More air conditioner will take care of it," Monkey shot back. "Ten people in here, I don't want something that don't work."

"Ten people?" I looked at Michael.

"We're putting in some more phones."

Apparently Monkey had landed a large new sheet of California customers on the heels of Bob getting the Chinatown business. To accommodate the increased volume we were hiring a couple of new clerks and adding four more phones.

"Where are we all going to sit?"

"We're setting up the four phones for the new sheet at that table," Bob said, pointing to Krause's teak dining table.

Spanky said to Monkey, "Is it true that if you have over a certain number of people in an office it goes from a misdemeanor to a felony?"

"Who told you that?"

"I don't know. I heard it somewhere."

Monkey said, "Ah, that's bullshit. Three, ten, fifty. It don't make no difference. Don't worry about it."

At the end of the session, Monkey started rearranging furniture. I helped him put the leaf in the teak table to make it bigger. Krause had come back out of his room to see what was going on. "This is crazy!" he yelled. "What are you doing with my table? It used to be two guys in a corner talking quietly. Now you're going to have a hundred people in here!"

Krause continued to rant until Monkey offered to pay him more money.

"I know that's what you want. You want some more cake," Monkey said.

"Go to hell," Krause said, moving into the kitchen.

Monkey said, "That's all he wants. More cake." But I knew that wasn't all Krause wanted. The money would just insulate him from what he wanted, make it less painful for him to deal with the fact that he didn't have the resolve to get what he really wanted. Which, in part, was a life that didn't include us.

As Michael, Bob, and I filed past the kitchen on our way out of the office, Krause looked up from a pan of something he was frying and shook his head sadly. "Harvard, Brown. What are you guys doing with your lives? I mean, really? Is this all you can think of?"

25

"CONNECTICUT?" Charlie said. "You're actually considering taking a job in Connecticut?"

"I'm not considering anything. I had an interview. That's all."

"What about our script?"

"Can't we deal with my job prospects when they're relevant?"

"They're relevant right now, Pedro. Shit. I thought you were committed to this screenplay."

"I am, but—"

"But?"

"Charlie, I'm trying to figure things out. I don't know if this Connecticut thing makes any sense. I'm just keeping my options open."

"Editor of an alternative newsweekly in East Bumfuck, Connecticut? I don't know, man. Is that what you've been building to?"

"Pal, I clearly haven't been building to anything. That's the problem."

"What's it pay? Forty?"

"I have no idea."

Charlie shook his head. We were at the dining table in his borrowed loft. The Chilean couple's cat, a neurotic calico named Rusty, trod across my yellow legal pad.

"Just tell me one thing. Is this some fantasy of yours that involves Anna?"

"No."

"Are you sure?" Charlie's eyes dared me to lie. I broke the contact.

"C'mon, we're wasting time," I said, sweeping Rusty off the table. "Let's get to work."

"Connecticut," Charlie said. "Christ."

26

I'D looked at several apartments in the Village. The pickings were slim in the $1,200-a-month range. I had six different real estate agents working for me, and they were all saying the same thing: for $1,400 I might be able to find something. Might. But that was no good. It meant I would have to lay out $ 1,000 a month myself. That was too much.

When I mentioned to Monkey that I'd been looking, he didn't seem startled at all. He said, "You know what to get."

I shrugged. "I think so."

"Just make sure it's someplace nobody's gonna pay attention to us. So we can come and go without anyone noticing. And there's parking."

I said, "It's the Village. It's not great parking."

He said, "Don't worry about it."

"I thought you needed parking."

"It don't matter. See what you can get."

"I'm still not actually a hundred percent certain on this."

"What's the problem? It's just something to help you out while you get back on your feet."

"Yeah, I know, but—"

"You don't wanna do it, don't do it. Makes no difference to me."

The studio in Brooklyn was mine for another five weeks, so I didn't have to rush into anything. I'd wait and see. Besides, there was Connecticut. Anna had told me, after I called to tell her about the interview, that she kind of liked the idea of living there.

I began to get a picture of a little house back in the woods with a pond nearby. Maybe a dog barking and running after my car. A big mutt, half golden retriever, half something else. The kind of dog that would make Anna happy.

27

"I didn't take your advice," Michael said.

"What advice?"

"About Shelly."

"Michael, I didn't give you advice."

"Well, I didn't take it anyway."

He proceeded to tell me about his evening at the Plaza Hotel with the girl of his dreams, in which he'd arranged for the bridal suite, several bottles of Cristal champagne, fresh strawberries, and a gram of cocaine.

"So it sounds as if it went pretty well."

"She moved out."

"What do you mean?"

"From Tuna's. She moved out."

"After one night with you?"

"What can I say?" He smiled and shrugged. "No, she'd been wanting to for months. She just hadn't worked up the nerve."

"She moved in with you after one night?"

"Not with me. With a girlfriend. But do you want to know the really funny part?"

"I'm not sure I do."

"I talked to Tuna this morning. He called to get his figure, and he told me about her moving out. He was all broken up. He's positive she has a new boyfriend." Michael laughed his constricted laugh. "He said, 'If I find out who the guy is, I swear I'll kill him.'"

"Jesus, Michael."

"What?"

"You better hope he doesn't find out."

"Ahh," Michael said, shrugging. Just to be safe he and Bob had arranged for a friend of theirs to start booking Tuna's bets. "I told Tuna we were having some trouble with the heat, so I was putting him in at another office. Of course Brent is just a friend of Bob's who's doing it for us as a favor—we'll funnel the money back and forth through him. At least this way, even if Tuna finds out about me and Shelly we have a shot at keeping his business."

"I think I'd be more worried about losing my health than Tuna's business."

"I'm worried about losing his business," Michael said.

Michael's dangerous love life wasn't the only hot news at the office. Somebody had tipped us that our outs in the Dominican Republic— one of the places we called to lay off bets—had been busted.

This didn't seem to concern anyone else much, but it gave me a distinctly queasy feeling. I hadn't even been aware that we had outs in the Dominican Republic.

"So who are these guys?" I asked.

No response.

"Is it a connected operation?"

I might as well have been talking to myself.

More and more, I was of the opinion that we were in bed with some heavy hitters.

I called Michael at home that evening to get the lowdown.

"It's nothing," he said.

"I just want to know about the real risks involved here."

"You know the risks. We could get arrested."

"Yeah, but if we're tied in with organized crime maybe it's worse."

"Look, when guys have been busted recently, the clerks have gotten 'promoting gambling' misdemeanors and five-hundred-dollar fines. That's it. Unless you get a prosecutor who wants to be a son of a bitch, which isn't likely."

I said, "But what about the over-a-certain-number-of-people-in-the-room-it's-a-felony law? Is there anything to that?"

He said, "I'm telling you. Last month the Nickel office got busted. There were ten guys there. They were out the next day and up and running with just misdemeanors." He sounded impatient with me.

"Listen, Michael, I'm just trying to sort out what the story is here. I'm hearing different things."

He said, "Well, if anyone should be worried, it's me. I'm at more of a risk than you. I'm in deeper."

28

CHARLIE and I kept running into problems with the story line.
Sometimes we'd sit at the dining table in his loft for long silent
minutes, each of us trying to figure out solutions.

The temptation, always, was to get up and take a break.

"Are you hungry?"

"I'm starved."

"You want to go get something to eat and figure this out later?"

"Chinatown?"

"Beautiful."

Frequently we'd just start talking and that was our break. Charlie
kept after me for details about my new job. He wanted to know if I
was betting. "With all these guys wagering, doesn't it make you want
to?"

I told him that in fact it did. "I know I shouldn't be doing it, but I'm
only betting fifty a pop. My code name is Curly for Topsider."

"Maybe I should give it a try."

"Save your money."

"You're losing?"

"A little."

"Don't you get inside info?"

"It's not as much fun that way. I like using my own opinion."

"When do you start making the kind of money Winnie's kid is making?"

"When we write this script, pal. C'mon, let's get back to work."

Charlie always wanted to put a twist on things to avoid predictability.
He'd say, "Let's make the hero a woman. And let's make her black."

I'd say, "That's not the problem. If we need to do that to make the story original, then there's something already wrong." And he'd go, "No, I think the guy should definitely be a black woman."

We could get pretty hot at each other. It was like a romantic relationship: inevitably we'd reach the point of thinking that maybe we just weren't right for each other. Maybe this wasn't going to work out.

But we kept bucking the doubt. We made headway.

Which was good because, although the Chilean couple wouldn't return for another month, Charlie was starting to think that for the sake of his marriage he might have to go back to L.A. pretty soon. We agreed that we needed a finished outline before he left.

29

I HADN'T given Anna my work number because, in theory, personal calls were not allowed at the office. But Monkey, Bob, Michael, Eddie—all the partners in the business—regularly disregarded their own rules.

Bob was the worst. He got calls from his girlfriend all the time. Monkey wasn't much better.

Nothing pissed off Pat more than someone getting a personal phone call. He never got any.

"Monkey," Pat said as he handed him the phone during a session, "tell your wife this is a place of business, not a fucking social club."

"And while you're at it ask her if she wants to meet me in the usual spot or should I just go right to your place?" Bob said.

Monkey rolled his eyes, slapping the phone to his ear. "Whaddja buy now?" he rasped. "Don't tell me. I don't want to know." He covered the mouthpiece. "If a day goes by that she don't buy something, the stores start calling the house to see if she's feeling okay." He lifted his hand. "No, I'm not talking about you. I'm at work. I work here. I make money so you can go out and spend it. . . . What? You're going where? With your sister?" He made a face. "Sure, I believe you. I'm going to come home and smell you tonight, see if you smell like sex. You better not smell like sex."

"Twenty years and they talk like they're still in love," Bob said.

"Yeah, talking is all he can do at this point," Pat said. "The man hasn't had a hard-on since 1972."

"That fucking kid," Monkey said, after he hung up. "Can you believe it? He got in trouble again."

Pat said, "What'd he do this time? Shoot a neighbor?"

"Nah. He cursed at his tutor."

They were talking about Monkey's eight-year-old son. "What's he getting tutored in?" Bob asked.

"Math."

"Math? The kid knows how to lay points and figure the vig and he's failing math?"

"It's different. He don't make no money in his math class, so he don't concentrate."

"Lucky for him he's got you as a role model," Pat said. "You teachin' him about broads, too?"

"Teachin' him? These kids know everything already."

"So he's got little eight-year-old hookers coming over after school?"

"I'm telling you. This kid ain't like me. He don't even have to pay."

30

A couple of weeks went by and I didn't hear anything from Con-necticut. I assumed that meant they weren't interested, but I tried calling anyway. I left messages.

I wasn't at all sure I wanted the job or that I'd have taken it, but I was at least expecting to get a follow-up interview. How could my take on the first interview have been that far off-base?

When I went into the office, I told Monkey that I was still looking for an apartment. Too late, he said. They'd already found a place.

So that was taken care of, too.

The new place was farther downtown, I was told. Down on Canal. We'd be moving shortly.

That got me thinking about the Chinese gangs, wondering if it'd be dangerous having the office right there. We kept a lot of money on hand sometimes and those Chinese gangs were crazy. I wondered if Monkey knew what he was doing.

Michael apparently didn't. After all the risks, the claims that she was worth it and that he intended to marry her, Michael dumped Shelly a scant three weeks after taking up with her. It didn't surprise me. I knew he was too smart for her, that she had to be a bimbo to have lived with Tuna for so long, but Michael had been too dazzled by her looks to see it right away.

When I asked him what happened, he said, "She was starting to get on my nerves, so I told her it was over. I don't know. I think I just don't like women."

Krause came out of the kitchen carrying a cast-iron frying pan with sausages still sizzling in it. "You hate your mother," he said. "That's what it is. You may not realize it, but you hate her. That's why you're doing this." Krause gestured around the room with the pan, as if to encompass not only Michael's treatment of women but his involvement in the bookie business.

I had a suspicion that Krause was right, and I wondered, Is that why I'm doing this? Do I secretly hate my mother? And is that what she'll think if she finds out?

31

Every Monday when I went into work there was an envelope with my name on it containing my pay in cash. I generally got paid in hundreds unless a customer had stuck the office with small bills, in which case I got stuck with them, too.

I wasn't making that much—about $400 a week—but it was nice having the cash on hand. It came in handy when I found a decent apartment about four blocks from where I was staying in Brooklyn Heights and had to act fast to nail down the deal. The apartment was in an old building with hallways that smelled of cabbage and apple pancakes. What the two rooms plus kitchen lacked in size, they made up for in character: dirty, scuffed, plaster-streaked wood floors, original moldings, tin ceilings, great light. I would have liked to delay signing the lease until I heard something definite from Connecticut, but I was too afraid I'd lose the place. So I signed and forked over a couple of months' rent in cash.

The move-in date was September 15. I wrote down the date in my Week-at-a-Glance calendar to commemorate it. One year of post-Chicago limbo, over. Flipping through the previous eight months, the mostly blank pages of the winter, I had an image of myself moving through the large empty house in Provincetown like a ghost, waiting for life to return, afraid that it wouldn't.

But I wondered: Was this the kind of life I'd had in mind?

32

I picked up a Snapple and a bag of Sun Chips at the bodega on the corner. August had arrived with a vengeance. It was 98 degrees out, a fact that both the *Post* and the *News* had duly noted on their front pages. The *Post* had a drooping 9 and a drooping 8 and a drooping degree sign; the *News* simply put a picture of a tongue-lolling pooch under the banner "Dog Daze."

Two uniformed beat cops were standing on the corner near Krause's building, talking to each other, one of them swinging his billy club. The sweat shone on their brows and ran in tiny streams down their cheeks. I tried for nonchalance as I passed them, feeling as if the criminality radiated off me like the heat off the sidewalk. They didn't even give me a second glance.

"What do you think this is, a fucking cafeteria?" Pat said as I dropped my brown paper bag on the worktable. It was just him, me, and Bob on for the day shift. Despite all the talk about new clerks and new customers, the crew hadn't changed since I'd begun working.

"You think I'm kidding?" Pat said. "This isn't a restaurant."

"He's just pissed because you didn't bring him anything," Bob said, seeing me flinch.

"You want me to go out and get you something?" I asked Pat.

"Don't get him anything," Bob said. "Let the lazy motherfucker get it himself."

"Big fuckin' shots, you and Michael," Pat said. "With your slant-eyed new customers. What are you gonna do when the Chinks owe you a quarter of a mil and tell you to fuck off?"

"I'm not worried. We got Jimmy D. behind us," Bob said.

"Fuck Jimmy D."

"Go ahead," said Bob. "I dare you to."

"That hump. I'd break him like a twig."

Bob bugged out his eyes for my benefit. Pat caught him doing it.

"You think I couldn't?" Pat said.

"Jimmy D.?"

"I'll take on him and you and Michael and whatever other five guys you want."

Bob burst out laughing.

"Go ahead," Pat said. "You and Michael think this shit is all a game. You'll find out the hard way someday."

He looked at me, and out of fear I nodded in agreement. "Everyone finds out eventually," he said.

Earlier in the day, at the studio, I'd received a call from Stewart Granitz, the publisher of the paper in Connecticut. It turned out he hadn't forgotten about me after all. Which was unbelievable, me just having signed a lease.

He said they'd offered the job to someone else, but then the guy had turned it down. I'd been right near the top of the list, which Granitz realized he should have told me, but things had been "a little crazy."

"Anyway, I'd like you to come in for a second interview," Granitz said. "The owner of the paper wants to meet you. I won't tell you this is just a formality. But he generally goes with what I've decided."

Granitz's words reverberated now: "He generally goes with what I've decided . . .. *what I've decided . . .*"

Pat said, "You gonna pick up the phone, or are you just going to listen to it ring?"

"Huh?" I swam up out of my daze and reached for the black plastic handset. Bob and Pat shook their heads.

"Hello?"

"Hi, you got a line yet?"

"Who is this?" I asked.

"Moon for Charlie."

Charlie. I had to speak to Charlie. What was I going to say to him? And what about my apartment? What about my lease?

"Yo, buddy, you there? You got a line?"

"Yeah, yeah. Who is this again?"

"Moon for Charlie."

"Yeah, hi, Moon. I got the Pirates twenty-nine. Expos pick. Marlins oh-eight. . ."

Granitz had said in our original interview that he wanted the paper to do hard-hitting investigative stories, and I'd been able to picture myself in a gray fabric-paneled cubicle, with sleeves rolled up, tie loosened, phone plastered to one ear, computer keyboard at my fingertips, working deep into the night on the story that would . . .

Would what? Scoop the *Times*? Solve my life?

"Pedro, I hate to disabuse you of your Ben Bradlee fantasy," Charlie chided me when I called him after the session. "But you're not going to be working for the *Washington Post.*"

Pat and Bob had left, and Krause had made a rare foray into the outside world. I was alone in the office, my feet propped up on the desk.

"Investigative pieces take time," Charlie went on. "Where are you going to find writers who'll put in that kind of time for no money?"

"I don't know. Maybe some rich, bored suburban housewife will turn out to be the next Seymour Hersh."

"You're going to work like a dog, be extremely frustrated, and have no time left to do any of your own writing."

"I know, but—"

"I'm not just saying this because of our script. I think you're trying to romanticize this job. I think you're going to be disappointed."

"I know it's not going to be anything earthshaking. In a weird way that's part of the appeal—"

"You just want to be normal."

"Yeah. Exactly. I just want to be normal. I want to have a life, a family, a house."

"I know. I want those things, too. I think about those things."

"So you know."

"But, Pedro, Connecticut? A giveaway newspaper? C'mon."

I sighed. One of the other phones was ringing. The session had been over for fifteen minutes. But the phones kept ringing. "These sick fucks will keep calling 'round the clock," Pat had said. "They don't know from office hours. That's why Krause goes nuts when we forget to turn off the ringers."

"But what do I tell this guy Granitz?" I asked Charlie. " 'Sorry, gotta say no. I've taken a job as a bookie instead'?"

"He'll understand."

"And what about my mom? That poor woman. What am I going to tell her?"

"She'll understand, too."

"And you think *I'm* the deluded one?"

If talking to Charlie was no help, the conversation that I began with Anna that evening was a disaster. I don't know what I expected—that she would say, "Yes, take the job! I'll be on the next plane to help find us a house"? The fact is, I did want her to say that, and when she couldn't, when she said it was too big a decision for her to make right now, I couldn't hide my disappointment, my anger.

We hadn't seen each other in nearly a year and I was asking her to move to Connecticut to live with me. From the start my relationship with her had been powered by a kind of mad impulsiveness: a chance meeting in a museum in the middle of Mexico; drop-everything moves, her to New York, then me to Chicago; intense psychodramas; wild sex. It felt like a betrayal that she would now want to make considered, logical decisions.

I tried to tell her something to that effect, but it came out bungled. She said the whole conversation had the feel to her of a setup. I said her remark about liking the idea of living in Connecticut smacked of the same. We gotty pissy with each other, then tried to make it up. Anna said she didn't want to fight. I said I didn't want to fight either.

"I want to keep what's good between us," Anna said.

"Me too."

"Then let's try to do that."

"Let's."

"Will you call me when you decide about that job?"

"Sure," I said.

But later I got angry again. I wanted my life to make sense. And after I called Stewart Granitz in the morning to tell him I'd been hired to write a screenplay and wouldn't be able to accept the job, I thought, Why should I tell Anna? Why should she know?

In a couple of weeks, maybe even a couple of days, I would forget what had made me so angry. I would get lonely again and start missing her. But right now I tried to use the anger to help me face the day. To launch myself into that "brief crack of light between darknesses" without thinking that the day would be meaningless if she wasn't part of it.

33

MONKEY dropped in during the middle of the afternoon session. He was wearing a light blue Italian suit and white loafers.

"Ooooh, baa-by," Bob said. "Where are *you* goin'?"

"Nowhere," Monkey said. "Take this and pay this guy for me." He handed Bob a fat rubber-banded roll of hundreds. "Three o'clock on the corner by the bagel shop. You know, over on First?"

"How much is here?"

"Ten dimes. Don't bother counting it."

Bob slid the money down inside the waistband of his boxers. Monkey asked Spanky if he wanted to take his shift for the night session.

"Can't. I have plans," Spanky said.

"Yeah?" Monkey said. "Since when do you got anything to do?"

"I have plans," Spanky repeated sullenly.

"Monkey's got plans too, right?" Bob said.

"None of your business."

"What's her name?"

"Names."

"Names? You tellin' me there's more than one?"

"Bullshit," Pat said, hanging up his phone. "He can't handle one woman, how's he gonna handle two?"

Monkey ignored him. "What about you?" he asked me. "You want to work tonight?"

"Hmm," I said, pursing my lips.

"What, are you guys so rich all of a sudden you don't need the extra cheese?" Monkey said.

"No, it's just that—"

"Hey, it makes no difference to me if you don't want the money."

I said, "All right, I'll do it."

"It's no favor to me," Monkey said. "I don't care one way or another. I'm just trying to help you out."

Taking the evening session meant that Charlie and I would have to push back our script meeting until after eight. Now that the job in Connecticut was no longer a possibility, the script was the only buffer between me and the reality of how I was actually spending my time. I was reluctant to lose a couple of hours with Charlie, since he'd be going back to L.A. soon. But I couldn't pass up the money.

Before he left, Monkey told us that we'd be moving into the new office at the start of football season. The newspapers were already full of training-camp stories. Preseason games kicked off the following weekend. It was still early August.

"I can't wait," Bob said. "Football can't start soon enough for me."

"I hope I still have two cents to rub together," Pat said. "Fuckin' baseball. National pastime my ass."

34

THINGS with Charlie fell apart a few days later. Ironically, the schism came less than twenty-four hours after we danced down Mott Street, celebrating our future.

What happened was this: On Friday night after work I met Charlie at the loft. He was tired after a day of running around town after editors, and I was tired from dealing with the macho pressures of the office. During dinner at a Thai place in Chinatown, we had a couple of beers and got on a roll. The movie came alive before us. I could see the characters; I knew them and cared about them; they were no longer constructions. Each twist and turn we took seemed both inevitable to me and surprising. By the time we wandered over to a sidewalk cafe in Little Italy that was either brave or stupid enough to be sporting a Dinkins for Mayor poster in its window, we were both excited.

Drinking one iced cappuccino after another, we hammered out the final few scenes and then Charlie rocked back in his white plastic chair, eyes brimming. "Pedro," he said, shaking his head, "we're gonna do it, man. We're going to be rich."

"All we have to do now is actually write it."

"But we got it, man. We really got it."

We high-fived each other across the table. Palm to palm, backhand to backhand, palm to palm. And then we got up and danced around while the couple who ran the cafe just looked at us as if we'd gone crazy.

Which in a way we had. Because the next morning Charlie called me and said, "Listen, I've been thinking. There's one thing that bothers me."

"Yeah?" My head felt thick, full of humid air and sleep.

"It's been bothering me for a while. I just don't believe the mistaken-identity part."

"What do you mean?"

"It's too much of a coincidence."

"What are you talking about, Charlie? That's the whole basis of the story."

"I know, but—"

"Are you serious?"

He was. He went on at length about it. I was flabbergasted. The doppelganger wasn't some minor detail of the story. It was the *hook* of the story, the element that had drawn me in to begin with, when Charlie had first talked it through.

"Charlie, this is insane," I said. "Do you know what you're doing?"

"I just—it's too coincidental. I have a problem with it."

I went bullshit. I knew I was overreacting, but I couldn't help myself. I had a familiar feeling in my gut. It was—well, Anna would have called it the feeling of having been set up. The euphoria of the previous night, the dancing down Mott Street and hugging each other—it was as if he were destroying that, purposely, willfully stomping on what we had accomplished.

More than anything, I felt betrayed. And in the worst possible way: in the way I had always betrayed myself—with a doubting, destructive voice that picked apart a work in progress and kept it from being realized. I thought, Shit, I don't need Charlie's help for this. I could do this myself. The whole point of us working together was to prevent this kind of thing.

I told him what I thought. He did his best to deflect me, saying, "Everything's fine." But I couldn't let go of my seething rage. For me, the trust was broken.

"Pedro, please, don't let this get to you, man. Please," he implored. "It's part of the process. That's all. We'll work it out."

"Yeah. I guess." But my heart was no longer in it. I was wary. He'd made me doubtful. We actually got together a couple of more times before Charlie went back to L.A. By then the delicate confidence that we could succeed, and that the script had merit, had been shattered. Officially, we decided to let things rest and see how we felt in a couple of weeks or months. Unofficially, I knew it was unlikely we'd ever return to it.

The day Charlie left, it was raining in New York but sunny in San Francisco. The Mets were playing the Giants in Candlestick, and they were a big underdog. The take-back on a hundred-dollar bet was $240. I wrote myself a $50 ticket using the name Curly for Topsider, and after work, instead of going back to Brooklyn, I stopped in a bar along Second Avenue that had the game going with the sound off, up high over one end of the bar. It was a close game through seven innings, though I knew if I hadn't been betting it would've held no interest for me. As it was, I kept shifting on my rickety wooden stool, eating stale, salty popcorn out of a dirty wicker basket, and looking around restlessly. There was one other guy at the bar, a few seats away, in baggy red shorts with a vein-blue tattoo running down his hairless calf. His elbow was planted on the bar and his head was bent forward and he kept running his hand through his crew cut, brooding over some private sorrow.

When the Giants scored four times in the bottom of the eighth, I decided I had seen enough. I paid my $7.50 bill with a ten and I left.

35

WE had two new clerks in the office, one of them a woman. Whatever was holding up the new California sheet had apparently been worked out, though of course no one had told me—I was too low on the food chain. I'd been assigned with the two new clerks to what we called the Rex table, named after the sheet's agent, where an 800 number had been installed to accommodate our out-of-state customers.

The woman clerk's name was Brandi, and she was an ex-girlfriend of Steak Knife's. Word was, he was trying to help her out because he still felt guilty about having dumped her. She looked like a pizza queen from Corona. Thirty-five years old, tight little body in skintight acid-washeds, bleached blond hair cut in one of those skunk-looking buzz tails, large tinted glasses, and translucent pink lipstick that probably had a name like Liquid Passion.

I thought the coarse atmosphere of the office might be softened by the presence of a woman, but aside from Pat, who was caught up in some grotesque parody of flirtation, there was no discernible change.

The other new clerk was one of Monkey's boyhood pals, Bernie. Bernie was about fifty but had the body of an old man. He smoked one cigarette after another, stopping only when his emphysemic breathing grew labored and he had to pull out an inhaler to clear his lungs.

Everyone was quickly convinced Bernie would die in the office, so there was a lot of gallows humor about how we'd dispose of the body. "Right out the window," Monkey said. "Maybe we'll put him in a Hefty bag first. Then out the window." Bernie rolled his eyes whenever the jokes started, jutted out his Gregory Corso lower lip so that you could see the wet inside of it, puffed out his cheeks, and lit up another cigarette.

Despite the fact that I was no longer the new kid, I remained the designated office gofer because Bernie was too infirm and Brandi was a woman. So I continued to plug meters and make sandwich and coffee runs. My only consolation was fantasizing that the cops would bust the place while I was gone. I'd come back and they'd be leading everyone off in handcuffs, and I'd pretend I didn't know them. And in that moment I wouldn't. They'd be criminals I was watching go off to jail. And I'd be a guy holding a bag of sandwiches.

36

MICHAEL and I were standing and talking on the corner near Krause's building when a truck backfired down the block.

Michael fell back against the brick wall as if he'd been shot. He looked quickly to the right and left, only getting his breath back when he realized no one had a gun.

"We're a little tense today, aren't we?" I said.

He shrugged, adjusting his T-shirt so the neck wasn't hanging off one shoulder. "Tuna found out about me and Shelly and threatened to kill me."

I looked at him the way his father might have if he weren't an alcoholic fuckup who had dropped out of his son's life.

"What?" he asked.

"What do you mean, 'What?' He threatened to *kill* you?"

"He's not going to kill me."

"Is that why you jumped five feet in the air?"

Michael shrugged. "He's not going to kill me. Shit, I'm not even seeing her anymore."

"Yeah, I'm sure he really cares about that."

I walked him over to the 57, a dive on Seventh Street between First and Second where he was a regular and king of the pool table.

"C'mon in, I'll buy you a drink," he said.

I had no plans. I was on my way back to the studio in Brooklyn, where I'd probably wind up watching TV. But I hesitated.

"Scared?"

"No, it's not that," I said, and then laughed when I realized what he meant.

"So you mean I should just be insulted."

"Fuck you. All right. Buy me a drink."

We went down a couple of steps. Inside, two ceiling fans spun lazily under the tin-paneled ceiling. Worn-out Naugahyde booths lined one side of the long narrow space; the bar ran along the opposite side; the pool table was in the back. The heavily made-up gray-haired woman behind the bar greeted Michael, as did several of the patrons. The prevailing style was chin fuzz and open oxford shirts, bustiers and nose rings. All of them, except the barmaid, were in their twenties.

Michael bought a bourbon and soda for me and a Budweiser for himself after first writing his name in chalk on the blackboard by the pool table. While he was there, one of the players whispered something in his ear and they talked for a while.

When he came back, Michael told me that the guy had warned him that Tuna had been in looking for him.

"What! When?" My eyes swept the room.

He waved it off. "Earlier."

"Fuck, Michael. Let's get out of here!"

"Nah."

"You're just going to stick around till he comes back?"

Michael shrugged. "He won't be back. It's all bullshit. He comes in here before eight when he knows I'm not going to be here. He knows I'm still going to be at work."

"So? He's stupid."

"He's grandstanding. He's acting tough because he doesn't want people to think he's a pussy."

"If that's what you really think, you better not go around advertising it. If he hears you're doing that, he might feel compelled to prove you wrong."

Michael picked up a *New York Post* from the bar and opened it. "He *is* a pussy. If he was going to do anything, he wouldn't have told anyone." He flipped a page. His face was calm in the smoky yellow-brown light.

"In that case you should definitely keep your mouth shut."

Michael closed the paper and tossed it aside. "The part I really love is that he's still betting with us and doesn't know it."

"He hasn't caught on yet?"

"Nope."

I found myself feeling parental again. "Don't get too cocky, Michael."

"Why not?"

"It's dangerous."

"How do you know?"

That hurt me more than he could have imagined. "I'm just telling you. Don't push your luck."

He smiled and looked at me like a kid who knew who was paying my rent.

37

I'D been able to put away several hundred bucks so far. Each time I stashed one or two hundred-dollar bills inside a copy of one of the many books lying around the studio—*To Have and Have Not*, pp. 100-101—I'd think of Roy Dillon and his money-filled clown paintings in *The Grifters*.

Yet I kept getting the feeling I wasn't taking it far enough. I wanted more. I knew I'd have to pay my dues—that was the way things worked. Even so, I was getting impatient. Why not get rich? Or at least earn enough money to afford some good therapy?

Bob and I were the last ones to straggle out of the office after the morning session. He had his gym bag slung over his shoulder as usual.

"You going to work out?" I asked.

"Very observant, Mr. Harvard Man."

I kissed my middle finger and blew him a fuck-you.

"Listen, you got a few minutes?" I asked.

He glanced at his watch. "I'm a very busy guy. What is it? You figure out how to do baseball totals yet? You want another lesson now that the season's nearly over?"

"C'mon, man, what are you talking about? I know how to do 'em."

"Good. Maybe you can teach me sometime."

"Seriously."

"What? You want to *talk*?"

"Yeah."

"I hope this isn't the kind of talk my girlfriend means when she says 'talk.' "

"Very funny."

Bob cackled. Nobody could say he didn't appreciate his own jokes.

"So *what*? You want to interview me again for your book?"

I grinned. "I've quit writing. I want to make money now."

"You're a funny guy, Pete. You think you can make money doing this?"

"There's a rumor going around to that effect."

"Don't believe it. All you'll get is headaches and heartache."

"I got those already. They can't get worse. I just feel that as long as I'm doing this I'd like to be able to reap some of the rewards."

"Listen to me. If you really want to earn, you have to bring in customers. That's the only way."

"Is that how you and Michael did it?"

"Michael and I got very lucky when we started out. You can't look at us as an example. Some situations came up and we were able to take advantage of them." He told me how they had made a bunch of money during football season, betting middles. A middle was when you were able to take both sides of a game and work the spread in your favor. For example, the line on the Giants versus the Redskins might open with the Giants favored by four. So if you took the Redskins plus four for $1,000 and then the line moved to two and a half, you could take the Giants minus two and a half for $1,000 and have a middle. If the game ended with the Giants winning by three points, you'd have won both sides of the bet and made $2,000. Anything else and you'd have lost only the $100 vigorish for the one losing bet. Basically you were taking a 20-1 shot for your money.

So Michael and Bob, helped by good information, had gotten on the right side of a lot of games early on, and then they had gotten lucky on top of that by betting and hitting middles after the line moved, which had given them their working capital and an opportunity to press.

"It's the same story as in any business," Bob said. "It takes money to make money. Now we've got the money."

Plus, because Bob was adept at bringing in new bettors—the real lifeblood of the office—Monkey and Eddie and Steak Knife had seen fit to give the two kids a piece of the big pie. It meant that they, like the others, were involved in two different markets: the games they were playing themselves, and a percentage of the action the office was booking. They were diversified.

"Once you get into this," Bob said, "you can find lots of different ways to make money." He said I'd probably noticed that the office rarely laid off bets. The reason for that was that they weren't afraid to take a position. "We're gambling, but it isn't as much of a gamble as it seems. Not over the long haul." He explained: If there were fourteen games and the office took sides in ten of them, all they had to do was go five and five to make the same juice they'd make if the books were balanced. But if the public went four and six, which was likely, since the public tended to be wrong more often than they were right, then the office would show a substantial profit beyond the vig. Over the course of a season, the profit could amount to hundreds of thousands of dollars.

"The bottom line is, though—unless you get incredibly lucky, like we did—you gotta get players to start earning."

"That's the only way?"

"That's the only way."

38

Pat and Brandi arrived at the office together one morning. Bob whispered to me, "Either he hit the lottery or they're fucking. I've never seen Paddy smile two days in a row."

But Pat wasn't able to sustain his dopey smiles and nice-guy delivery even through the end of the week. A couple of days later he was back to his usual sour, acerbic self. In fact, worse. And I took this to mean whatever nascent romance had been in the air hadn't amounted to anything. Pat had the sulky, petulant demeanor of a scorned suitor.

Whenever Brandi opened her mouth, he scowled. Which was frequently because she was a nonstop talker. I found her friendly but annoying. I listened half attentively, not wanting to encourage her too much. It didn't matter. She was one of those people who were not deterred by a less than attentive audience. Her upbeat manner was relentless, her delivery sweetened by her own laughter. I found out she'd led "a crazy single life" for a very long time. Now she was raising her dead sister's eight-year-old. She'd gotten him five years ago and her whole world had changed. But he was a great kid. She owed him everything. He'd probably saved her life. "I was living way too fast," she told me.

I looked over at Pat, who mimicked her nonstop talking with his hand, moving it, fingers to thumb, like he was clicking a castanet.

"My sister had the Big C," Brandi was saying. "I guess it's a hereditary thing. Rollie's got it, too."

"Wait... the kid has cancer?"

"I still try to treat him like a normal kid. He hates being treated special."

"That's tough. Poor guy."

"The doctors don't know how long he's got. We're planning this big trip for the winter. He won it through a cancer foundation. He got to choose anywhere he wanted to go. They pay for airfare and two weeks' accommodations. He chose Australia because he loves *The Rescuers Down Under*. We're going in January."

The phones started ringing, and we had to work for a while. When I got off my line, Brandi was still on hers. I began reading the paper. This in no way deflected her when she finished her call.

"At the Tubby office," she said, "the phone never stopped ringing. From the minute you walked in to the minute you walked out, it never stopped. I liked it. It made the time go by fast."

"I can think of some other ways to make the time go by fast, Mama," Monkey said, leering.

Everybody laughed, including me, but I also raised my eyebrows afterward to let her know she had my sympathy.

She said, "Don't worry, Pete. I'm no virgin."

"Lemme ask you something, in that case," Bob said. "Do you like big penises?"

"Sure."

"See?" Bob said. "What have I been saying?"

"You didn't let me finish," Brandi said. "I was also going to say that a big penis isn't everything."

"Oh, sure, now you say it."

"Why are you so concerned about it anyway?" Brandi asked Bob.

"Because my penis is very small," Bob said.

Everyone cracked up.

"You're funny," Brandi said. "That's much sexier."

"Is that why my girlfriend laughs every time I undress?" Bob mimed a rim shot. More laughter.

"C'mere, Mama," said Monkey. "Sit here on my lap."

"Do you think Monkey is cute?" asked Bob. "Women always seem to think he's cute."

"He's got smiling eyes," Brandi said.

"Does that mean he's cute?"

"It means 'c'mere,' " Monkey said. He waved her toward his lap.

Giggling, Brandi went and sat on his lap. Monkey stuck out his tongue and rolled his eyes, bouncing her on his knee. Pat was talking to a customer on the lead phone, but Monkey's phone rang a couple of times unanswered.

"What is this?" Pat put his palm over the mouthpiece of the phone. "Is this a fucking bookmaking office or high school? How many times is that goddamn phone going to ring before someone answers it?" His face was thermometer red and rising.

Spanky tried not to lose it as Pat went back to talking to his customer. Monkey nodded Brandi off his lap and picked up his phone. He looked sternly at us, shaking his head, a signal not to laugh.

Spanky's phone rang. My phone rang. We looked at each other and did all we could not to explode. One word, one sound, was all it would have taken.

One word, one sound. It would've been all over.

39

WHEN I hung up the phone my heart was doing the double Dutch. Anna had crossed me up again.

After our previous conversation and from her tone at the outset of this one, I had assumed the worst was coming, especially when she told me that all of our talking back and forth had been upsetting to her. I thought, Oh-oh, man. Brace yourself.

Instead of a gentle letdown, she dropped a bombshell: she'd decided we needed to see each other.

I was so startled I couldn't speak. Typically she took my silence the wrong way.

"So what, do you hate me now because I couldn't tell you I'd move to Connecticut?"

"No, I'm just stunned."

"We don't have to see each other, you know."

"Did I say that?"

"You haven't said anything."

"Jeez, give me a break."

"Do you want to see me?"

"You really have to ask?"

"Well, I don't know."

"Of course. Of course I want to see you."

After we hung up, I sat by the window for a while and vibrated. I could see the planes coming in over the city. It was Tuesday. She was flying in on Friday. Maybe I'd just sit there until the plane she was on came into view.

40

TUNA got arrested. He and two associates broke into a townhouse on the Upper East Side over the weekend and tripped a silent alarm that was wired directly to the Nineteenth Precinct. He's facing a six-to-eight-year term for breaking and entering.

You'd think that Michael would have been delighted by this news, but instead he claimed to be considering putting up Tuna's bail.

"He was our best customer," Michael said. "I don't want to lose him."

I was incredulous. "The guy just got busted for a felony. You still think he wouldn't have made good on his threats against you?"

Michael maintained that he wouldn't have. Tuna might be a thief, but he wasn't violent. "I mean, look at us, we're breaking the law. That doesn't mean we'd hurt anybody."

"I'd be hoping they'd lock the guy up forever," I said.

"Nah."

"You keep saying all you care about is the money. If that were really true you wouldn't have risked sleeping with Shelly in the first place."

Michael pondered that for a while, studying me. I always felt that I was being judged when I talked to him, even when I was the one doing the judging.

Later on, my inexplicable need to impress him emerged in an even odder way: I told him that Anna was coming. He knew some of our history and said, "You're just a glutton for punishment, aren't you?"

I could scarcely deny it. "At least I'm not bailing her out of jail," I managed.

41

SHE came in on United flight 53 at 9:35 Friday night, and I was there at the airport to meet her.

She couldn't even look at me at first. Every time she tried, a sheepish, embarrassed smile would form on her wide mouth and she'd turn away. She was stiff and nervous. I was, too. "Overstimulated" was how she described it later.

We took a cab back to the city, Anna chattering away nervously, still without looking at me. I put my hand on hers and tried to penetrate the wall of words.

"I'm nervous," she said. "Can you tell?"

I held her hand, smiling.

"I always get like this, don't I?"

It was true, but I was glad of it, glad that the energy between us hadn't gone away. Suddenly my feelings, so carefully contained, seemed on the verge of exploding. I loved the way her tangle of dark blond hair spilled out from under her White Sox cap and how, in the moving grids of light that passed through the cab, I could see the tiny freckles around her nose and, briefly, before she looked away, her large green eyes.

"You really look great," I said.

"I do?" Her voice came out like a squeak.

"Yeah, you do."

I could see her taking it in, squirming a little but relaxing.

"Your hair is short," she said.

"I got it cut yesterday."

"Because you know I like it that way?"

"Why do you think?"

She squeezed my hand. "At least I didn't cut all mine off this time."

We both smiled. The day before I'd moved to Chicago to live with her, she'd hacked all her hair off with a pair of meat-carving shears, an act she rationalized as the least self-destructive course she could have taken.

It had grown back now to the length it was when we first met. When I pointed that out, she got a sad look in her eyes and said, "Oh, Pete, I wish—" but then stopped short.

"It doesn't matter to me," I said, "all the stuff that's happened."

"Yeah . .." she said. But I knew it wasn't that simple to her.

"How's Nathaniel?" I asked.

"I left him with his father. He's good."

"Does he ever talk about me?"

"Honestly?" she asked, looking into my eyes for more than a brief glance. "Not anymore." She touched my arm. "You know how kids are."

In Brooklyn we climbed up the four flights, the worn green-carpeted stairs creaking under our weight. I had worried about how things would go when we were in the close quarters of the studio, her straight off the plane, without the buffer of a full day together.

I probably should have just jumped her. But instead we opened a bottle of wine and tried to pretend we weren't nervous. When it became obvious that wasn't working, Anna put her wine glass down and said, "This is dumb. Let's just go to bed."

We sat on opposite edges of the bed, taking our socks and shoes off like an old married couple, her, in panties and T-shirt, climbing under the sheet on one side and me, in blue Jockeys, slipping in on the other. We turned, a little shyly, to face each other.

"This reminds me of that first night in Mexico," I said, thinking of the huge bed she'd had and how we'd gotten in without ever having touched or kissed before, and how we'd talked for half an hour, face to face, until I'd finally reached over and run my fingers along the underside of her wrist.

I did that now, moving closer and kissing her, gently, without using my tongue.

"I feel shy with you," she said, drawing back, giggling.

"It's okay," I said. "We can just sleep if you want."

"I don't know why I feel so shy."

"It's just me."

"It is you, isn't it?"

She moved closer, draping a leg over my thighs. I felt myself growing hard, and I started kissing her again. I lifted her T-shirt up and off, planting kisses all over her warm skin, her breasts, her neck, back up to her mouth.

It was okay, she said, for us to make love with only her diaphragm.

"What d'you mean?"

"I mean it's okay. I've been tested. Samuel got tested, too."

I felt myself go rigid in her arms.

"I'm sorry," she said. "I know you didn't want to think about that."

I hadn't thought about it. I had done a really good job of not thinking about it. The fact that she had been sleeping with someone else since January.

"It's just there, Pete. We might as well deal with it."

"I know."

But we didn't. Not in words, anyway. Instead, she rolled her weight on top of me, grabbed my face between her hands, and kissed me. She drew back, then came in again from a different angle. She did this again and again, more fervently each time. I reached down; she was soaking wet. "Let me feel you," she said, taking my hand away.

She rode me slowly, struggling to find a rhythm. Looking in her eyes I couldn't seem to catch her. After a while I turned her over and we tried with me on top. It was better. But still not great. We were both thinking about him.

It was unnerving to be out of sync; bed was the one place where doubt had never snuck up on us, the one place where we were always sure.

All I could think was I had to fuck him out of her system. That was the only way we'd both get past it. We both knew. So we went at it again, desperate, angry, trying to strip something away and get something back all at once.

Somehow it worked. Somehow, there in the early morning, as first light began to make the blinds glow, we found each other. I felt that she was mine again. Her arms were tight around my neck. She didn't want to let go. I breathed in her warmth, the smell of her sweat, the clean scent of her hair. Anna was the first to be made nervous by how good it felt. "I don't want this to throw off my equilibrium, Pete," she whispered. "I don't want to be figuring out everything in terms of this."

"I know."

"It's unsettling."

"But you don't want it settled. You hate settled."

She smiled at me slowly, looking straight into my eyes. "You know me pretty well, don't you?"

I nodded.

"And you think I should just be cool about things."

I nodded again.

She moved closer to me until I could feel her breath against my lips, feel her weight and flesh. The blinds were drawn, but the window was open and the plastic slats swayed in the breeze off the river. We stayed cool.

42

No one said boo when I came in. Spanky was cleaning tapes with a large electric magnet that resembled a steam iron. It made a cheap sci-fi buzzing sound when he pulled the trigger. *Bzz, bzz.* Pat was clicking his sports beeper, checking scores from the day games. Monkey was in the bathroom taking a piss with the door wide open. The a.c. was going, but it had obviously been turned on only minutes before, because the room was stifling. I went and stood right next to it to try to stem the sweat I'd worked up on the walk from the subway. Michael glanced at me over the top of his *Post.*

"What are you all dressed up for?" he asked.

"Anna and I are going out later."

"Oh, right. She got here."

"Your girlfriend's in town?" Pat asked.

"You got a broad?" Monkey's voice rumbled from the bathroom.

"Yeah."

Michael stretched, half yawning. "So did you get laid?"

I laughed. "Come on."

"Did you?"

"What do you think?"

"What's it been? A year?"

"No." I made a face.

"I thought you said you hadn't been with any girls since you left Chicago."

"It's only been nine months."

"Oh, only nine months. That's different."

I gave him the finger.

"Where you taking her?"

"That place, Jules, up the block. I promised I'd take her to Paris. That's the nearest I can come on what you guys pay me."

"What time?"

"Right after work. You want to say hello?"

"Yeah, maybe."

I don't know if it was because my thoughts were on Anna, or if it was lack of sleep or what, but I kept messing up during the session.

I wrote a ticket for a player whose name nobody had ever heard of; I missed a couple of line changes; I wrote down a bet for a team with another team's betting line so that it was impossible to know which team the bettor had wanted.

By eight o'clock I'd had to eat a lot of shit. More shit than usual. The longer I worked at the office, the more comfortable everyone got ragging me. My only means of protection was to blow it back. But I wasn't used to doing that, and I found it even harder on a night when I was screwing up.

Michael walked with me over to Jules afterward, and talked to me like my job might be in jeopardy.

"You can't keep fucking up," he said. "It's costing us money."

"I had a bad day. Give me a break."

"Mistakes are expensive."

Anna was standing on the sidewalk outside the fenced-in outdoor terrace of Jules, by a mock guillotine that a photographer had set up for tourist photos. Anna had on a short floral-print cotton dress and Capezios and red lipstick. I gave her a kiss. "You remember Michael, don't you?"

They said hello, and the three of us stood there and made small talk for a few minutes. After Michael went on his way, Anna said, "He's a little stiff, isn't he?"

"Yeah, he is. He's all caught up in this money thing."

"Is that what's going to happen to you?"

"I may not last that long," I said, and told her what had taken place at the office.

She frowned. "It sounds like a really unpleasant atmosphere to work in."

"It's a bunch of very tough guys, you know what I'm talking about?" I poked at the air with my index finger. "I fit right in." Anna looked at me. And with no discernible irony, given the fact that she knew I was a neurotic Jew with back problems and food allergies, said, "Are you gonna end up being a tough guy?"

"I'm already a tough guy," I said. "C'mon, I'm starving." I took her hand and led her down the steps.

43

"You bastard," Anna said. "You're making me fall in love with you all over again." We were walking along Pierrepont—which, much to my amusement, she pronounced Pee-yare-pon—headed toward the subway and Manhattan. It was a crisp blue late-summer morning with some autumn in it. Throwing her arms around my neck, Anna clung to me so that we tottered crazily to and fro like a couple of drunks. Passersby stared at us with knowing little smiles. I tried for cool nonchalance, as if a beautiful girl clinging to me was my due, but inside I was grinning from ear to ear. The self-protective armor I'd strapped on before the weekend was gone. I was gone.

We met my friend Ezra for brunch at Les Deux Gamins, a small cafe near Sheridan Square. After two days of floating around New York, eating, drinking, making love, going to movies *(Map of the Human Heart* at the Angelika, with Anna sitting in the row behind me because there were no two seats together, her hands on my shoulders the whole time), and hearing music (a Caribbean funk band at S.O.B.'s that we danced our asses off to), it seemed important that we have a witness.

Ezra was perfect because he was among the least judgmental of my friends. He'd met Anna before and liked her, and he charmed her during lunch by asking her about her Louisiana roots (it turned out that his wife was from the same area as Anna's grandmother). After lunch, Ezra went on his way, and Anna and I sat by the fountain at Washington Square Park, watching the kids frolic in the dirty water while all around us bikers, skateboarders, stand-up comics, jugglers, and drug dealers practiced their craft.

Anna pushed her Ray-Bans up on her head. "I like Ezra," she said. "You haven't told him all sorts of horrible things about me, have you?"

"No, of course not," I said, feigning outrage.

"You promise?"

"I promise."

"Because you know you've done that in the past," she said.

It was true. In describing our troubles to my friends I had sometimes divulged certain facts about Anna's childhood that made her feel exposed and betrayed. It didn't matter that I might have been trying to explain to my friends—and to myself—why she sometimes got so depressed or acted so crazy. From her perspective, I had broken a trust, nothing less. Whatever she might have done to me in her unhappiness couldn't possibly have been equal to that.

"I promise you I didn't tell him anything," I said.

All of a sudden there was a funny tension between us. The end of the weekend was looming.

The next morning, lazing around in bed, Anna told me she might take a job in Bolivia.

I thought she was joking. "Bolivia?"

"You could come."

"C'mon, Anna."

"It's just for two years. It might not even happen."

"I can't believe this. You drop this on me now?"

"I thought I'd better tell you."

"Jesus!" I drew my hand out from under her and pushed my head back into the pillow up against the wall.

"If the grant money for the project comes through, we'd go in November. Isaacs wants me to be the coordinating director of the project. Thirty thousand a year in Bolivia is like a hundred thousand here. You could come and write."

"But Bolivia? Christ, I don't know anybody in Bolivia. I don't even speak the language ..."

"If the money comes through," she said, "I'm gone."

I was furious. No doubt all she wanted was for me to give her my blessing, tell her, "Go. That's great! What a fantastic opportunity."

She might genuinely have wanted me to go with her. But all I could think was that I'd been duped. She'd sucked me in, then laid out her escape plan and asked me to approve it.

By dinnertime we'd stopped talking about the Bolivian plan. But the frost that had replaced outright hostility was even more unsettling. Anna was going back to Chicago the next day, and we'd simplified whatever questions her departure raised by wiring the bridges to explode in her wake.

Her plane was to take off at six in the morning, which meant she'd have to leave for the airport at four. I got into bed around one. She took a seat, still fully dressed, in the red velvet chair by the window. She pulled a book off the shelf, turned on a small halogen lamp, and began reading. Would the light bother me?

"No, but don't you want to get some sleep?"

"I'm not tired."

Was it that, or could she just not stand the thought of lying next to me?

At a quarter to three I was still wide awake. I looked over at her. The light was burning. She heard me rustle but didn't look up.

"This is so stupid," I said.

She pretended not to hear me.

"Anna, just come over here."

She put down her book. Still, she didn't move.

I drew the sheet aside and got up. I walked to the chair, which sat in a pool of light, and stood over her, staring.

"This is crazy," I said.

She finally looked up, eyes flat.

"I don't understand this," I said. "How can things get like this with us? How does this happen?"

Still no response.

"Things were so great for a few days and then ... I don't know. It just goes wrong. It—"

She lashed out at me with a foot, kicking me in the thigh with her smart rubber-soled shoe.

"Jesus!" I was stunned.

"I hate you!" She kicked me again, hitting me in nearly the same spot, missing my balls by inches. "God, I hate you!"

She looked as if she were going to kick me again. I grabbed her wrists and wrestled her to the floor, pinning her down. "You fucking kicked me!" I screamed. "You kicked me!"

She struggled to move, twisting her head from side to side, making frustrated grunts and whimpers, trying to kick me the way Rosa Kleb tried to nail James Bond with her stiletto-toed shoes in *From Russia with Love.* "Let go of me! Let go!"

"That's it!" I yelled, keeping a firm grip on her. "This is so insane! I can't even believe this is happening."

"You're hurting me."

"You fucking kicked me!"

"You're hurting me."

"I don't care," I said. "I've had it. I want you to get the fuck out of here." I let her go and stood up. "Where's your bag?" I didn't wait for her to answer. The bag was already packed and sitting to the right of the door, by the bookshelves. "What else do you have, anything?"

She shook her head.

"Then get up. I want you out of here."

"Pete—"

"I don't want to hear it. Just get up."

She rose to her feet and moved toward me, looking positively sorrowful. "Please don't make me leave like this."

I waved her off.

"Pete, please."

I opened the door and waited for her to go. When she didn't move, I put my hand in the small of her back and half guided, half pushed her out of the studio. "All right," she said. "All right."

I hustled her down the stairs to the vestibule. Through the glass windows of the double doors, the street was dark. The taxi she'd arranged wasn't due for an hour. I didn't care. I was wearing only my Jockeys and didn't have my keys. "You can stay down here or go outside, it's up to you. I don't want to have to look at you or talk to you anymore."

"But—"

I threw up my hands as if I couldn't bear to listen, which in fact I couldn't, and without saying any more I climbed back up the stairs.

Twenty minutes later there was a soft knock on my door. I'd been lying in bed, running the whole episode through my head again and again, already feeling my anger fade and depression take its place.

I padded barefoot across the rug and pulled open the door. She stood there, looking sheepish, a kind of loopy grin working its way across her lips.

"Do you think what just happened might have had anything to do with the fact that I was about to leave?"

I stared at her for a full twenty seconds, then started to laugh. We were so obvious, so predictable, there was almost nothing to do but laugh.

"Do you know how fucked up we are?" I asked.

She shrugged. "It's not just me, you know."

"I know. Believe me I know."

I stepped aside to let her back in. She picked up her bag and entered.

44

TUESDAY night was the College Kickoff Classic. The office was hopping, a preview of what it was going to be like for the next five months, except that everyone was saying, "This is nothing. Wait until the full slate tomorrow and then Sunday with the pros."

Whenever the wise guys—Needle, Laser, or Panther—called, the person answering the phone was supposed to knock on the table, bringing the room to an E. F. Hutton halt. Then, whatever the play was, we would immediately jump the line and start dialing other bookies to get a bet down in the same direction. Within a minute all the followers would call up asking about that particular game. It was amazing. The same thing had happened in baseball, but for some reason in football it was more intense, more important. The sharp guys were sharper, and there were more followers. News traveled fast in the gambling world. Hot sides—which was what the wise guys' plays were called—didn't stay hot for long. Within half an hour, in every office from California to Maine, the line got corrected.

Needle was the code name for Billy Walters, one of the biggest and sharpest gamblers in the country. Laser was the code name for Danny D., a New York boy who was also a sharpie. What made them sharp? Track record. They'd proven over time that they knew what they were doing. It was like the stock market. Certain analysts have an uncanny knack for picking winners and find themselves with legions of followers, all of whom are trying to be the first to jump on their picks. Our deal with the wise guys was that we were happy to accept up to three thousand dollars per play from them, but only if we were part of their first wave (the really big bettors had offices of their own, with clerks who were given a buy order at a certain number and then would call out to ten bookies simultaneously). By being among the first they called, we had a chance to adjust our line and bet out with other bookies before they moved *their* lines.

After work, Pat asked me if I wanted to go to a poker game in the Village. The timing was perfect, with Anna gone and me not wanting to go home and brood. I went with him. Getting into his shiny black Eagle, I said, "Nice car," and he said, "What? This piece of shit? I'm getting rid of it as soon as I can. I'm getting a Lexus."

On the drive crosstown, just making small talk, I asked him about himself, and he told me about his job at No Excuses jeans, how he'd been brought in to trim the fat—they had all sorts of excess, jets and boats and unnecessary divisions—and how he'd done the job but then, for reasons that were complicated, gotten fired.

Apparently the same thing had happened to him at a couple of other companies, and after the last time, as if he needed any more bad luck, his marriage broke up and Black Monday wiped him out. Lost his family, lost his job. Next stop: the wrong side of the law.

I could see that underneath his gruff, cynical exterior Pat was just another poor lonely bastard hungry for love. He was pissed off most of the time because he was convinced he'd been ripped off by life, but he wasn't willing to admit that it might be something other than bad luck that had brought him down, or that he might in any way be responsible. It was easier for him to see himself as a victim. It had become a self-fulfilling role.

The poker game was in the basement of Our Lady of Pompeii Church over on Carmine Street. It was a three- and five-dollar-limit game, cash on the table, with weird crisscross and multiple-card abominations that favored luck over skill. I wound up sitting at a different table than Pat, next to a friendly chucklehead named Artie Dowd who played nearly every hand. The guy, as Pat would have said, "had no clue." A perfect sucker. At one point he pulled out his wallet and he had a fantastic array of rewards cards from every casino in Atlantic City plus about a grand in cash. So I started talking to him about sports betting. Asked him if he had a bookie. Told him that was what I did. He gave me his phone and beeper numbers. I didn't tell Pat because I worried he might think I was infringing on his turf, since he had taken me to the game, but after discussing it with Bob (giving him Artie's bona fides—i.e., he was a contractor who'd occasionally bet with a bookie named Tony on Sullivan Street), I decided to take a chance. I gave him the code name Tool for Topsider, and my deal with Bob and Michael was that I'd get ten percent of anything he lost. But Bob also let me know that I'd be one hundred percent responsible for any debt that Artie didn't cover. I said fine. Whatever nervousness that thought caused me was more than outweighed by the image of Artie's wallet, the thick wad of bills and how easily they had come flying out.

45

Saturday morning, early on, the recitation of the betting lines was like a religious chant, a rhythmic dirge in the church of college football: "Fifteen and a half, thirty-three, four, a half, twenty-one, ten, six, sixteen, six, four and a half, one and a half, seven, thirteen .. ." The numbers came out of nine mouths at once, consonant, running together, punctuated by the brief ring of a phone that was plucked off its cradle almost instantly. Then the second wave of calls came in, the electronic chirruping was nearly constant: as soon as you hung up, the phone rang again. People began to bet. Pencils scribbled away on betting slips. A tip broke, the pencil was tossed, another one was picked up. "I'm sorry, sir, you wanted who? UCLA minus seven a hundred and fifty times. Okay. Anything else?"

We wrote, by my estimate, over a million dollars' worth of business on the opening day of the college football season. The stack of slips at the end of the session was eight inches high. That amounts to $50,000 in vigorish, but we did better than that, making somewhere in the neighborhood of $100,000, according to Michael.

Throughout the whole day everybody kept saying, "If you think this is crazy, wait till tomorrow."

Well, Sunday wasn't as lunatic as predicted, but there was definitely some adrenaline flowing and there was one big surprise: at 11:45 a.m. the famous Steak Knife turned up, startling everybody. It was my first encounter with him, and I was shocked mainly by how mild-mannered he looked. Clark Kent as an accountant. He was also, it turned out (though no one had mentioned this fact to me), Eddie's brother. His real name was Alex, and basically he looked like a healthier, nondegenerate version of Eddie, with Sears Roebuck clothes and a nerdy haircut. A nice Jewish boy. He sat around and kibitzed for half an hour before it got crazy busy, and then he left. As soon as the door closed behind him, everybody, Eddie and Monkey included, expressed surprise that he'd shown up and put himself at risk. "He must be pretty fucking bored," Monkey said.

Steak Knife also did something that was apparently out of character. He had a reputation for being extremely cheap, but he gave Bob a sports ticker as a gift. Monkey was shocked. "He gave you that? I don't believe it. He wouldn't give someone a piece of ice in the middle of winter."

When it turned out that the gift beeper was half an hour behind the rest of the beepers, Monkey said, "Call him. Tell him thanks for nothing. Thanks for the piece of ice."

By three-thirty the results from the early games looked bad. The Giants game was on TV, and Eddie was going nuts watching it. "These motherfuckers! All they gotta do is sit on the ball and that nigger carries it with one hand. What the fuck."

If the agony of watching wasn't bad enough, he was also punching at his sports ticker every two minutes so he could groan over the results there, too. Everybody in the office was swearing and making racist comments, growing more and more depressed.

By four-thirty all the profits from Saturday had been wiped out, plus. When one of the phones broke, Monkey went ape-shit trying to fix it. He was maniacal. There we were, down thousands, and all he could focus on was the phone. He took it apart, shook it, decided it just needed new batteries, and sent me out to get them. When he put the batteries in, he still couldn't get a dial tone. "We can't afford this shit," he muttered, and flung the phone at the wall. It apparently didn't break into enough pieces to suit him, so he went over and stomped on what was left of it.

Eddie was even more bummed, because in addition to what the office lost, he had personally dropped about thirty thousand over the weekend betting the wise guys' games. Earlier in the week, like Pat, he'd been talking about buying a Lexus. Now Michael said to him, "Eddie, what kind of car did you say you were getting? A Yugo?"

46

Wigstock. Labor Day. End of the long hot summer. In between shifts I went out into the madness of Tompkins Square Park to eyeball the nine zillion made-up queens and wigged-out freaks who had taken over the park for the day along with RuPaul and Deelite and whoever else had a flamboyant bone in their body.

What a world! I mean just who precisely was it made for? A hairy-chested queen in pink chiffon with a two-foot-high flaming orange Lucille Ball wig? A serpent-haired Rasta naked but for a loincloth and a spliff? Ring-nosed punks with death-white faces and upturned palms? Or was it made for Pat or Monkey or Eddie? Or me? I mean, who was happier? Who was having more fun?

Back in the office I tried watching the U.S. Open tennis on TV. But Krause wandered in from his cave, demanding attention. I wound up talking to him for over an hour. Mostly it was him talking to me, or at me. An audience. Him with his T-shirt and grubby Jockey air of pathos. He told me about Howie, the bookie who'd used his apartment before we did, when it'd been just Howie and another guy at a table in the corner. "Now it's completely out of hand, with all of you turds. That's why I'm asking Monkey for more money."

Krause told me about how in the past he'd been sent out to California a couple of times to deliver money. One time to Gabe Kaplan's house. "He was very nice to me," Krause said. "He offered to let me stay over, but I decided to take the red-eye and charge Howie and them the higher fare."

Soon Krause launched into a monologue about his dear departed dog, Suzy. How smart she was and how pretty. He could have been talking about an old girlfriend. "I threw her a toy once," he said, his voice full of rapturous pain, "and she sprang off this settee to get it. It was where she always sat. On her way back, she moved the settee back against the wall with her paw. It had moved when she'd jumped. Isn't that incredible? *She knew where it belonged!* She was so smart. Everyone told me how smart and pretty she was. Everyone." As if not convinced his words were sufficient testimony, he went into the other room and got his wallet. There was a picture of him, young and handsome, on one side and of Suzy, a golden retriever mutt, on the other. "Wasn't she pretty?" he said. "Wasn't she?"

I found myself looking at him—the only son of a Jewish mom, likely to die alone having done nothing with his life—and I got the heebies. I realized why everyone in the office had contempt for him: he was their worst nightmare of themselves.

47

SATURDAY night after work I took the next day's *Times* home. I had continued to read through the classifieds and make job inquiries over the phone, but it was mostly a through-the-motions thing, an easing of my conscience. One person I'd gotten friendly with, an editor at *Vogue* name Sarah Shapiro, had actually made it her business to find me a job, in the naive belief that I really wanted one. She was one of those supersmart, energetic people who liked to get in there, roll up their sleeves, and attack a problem. She was full of ideas which she would usually leave like notes on my answering machine: "I hear there's an opening at *Harper's*. Do you know Lapham? I know someone who might be able to introduce you..."

Of course, when Anna got wind of Sarah's interest in me (all right, when I *told* her), she was no more swayed by talk of *Harper's* or Lewis Lapham than I was, though for predictably different reasons. Why, she wanted to know, was this woman so interested in me? How long had she been conducting this job search on my behalf? And just why, exactly, had I brought it up, unless to arouse her jealousy?

We didn't get into a fight over it, exactly. Which in a way I wish we had. Instead it was, Look, it's fine. We're not attached. We don't live in the same place. You're there and I'm here and we don't have a plan, so let's be realistic. It was Anna in hyperrational mode. Cool and detached. And it gave me a sinking feeling. She wasn't saying anything we both didn't know, but it made me extremely tired to hear her say it. I could feel myself go flatter than an old Coke. We'd fucked things up, and now neither of us had the energy or the trust or the wherewithal or whatever to put it all back together.

So it was, Yeah, right, better be realistic. We were grownups; this was the way things worked. Or didn't work.

And I thought, When am I ever going to drop my childish romantic notions? When am I going to stop saying to myself, "But I don't want it to be this way"?

48

Eddie came in one day with a brown paper bag dripping lemonade. I said, "Eddie, you're dripping." He ignored me and kept dripping right on the table. The corners of his mouth and his mustache were white with either confectioners' sugar or cocaine. His glasses were in full-tint mode. I said it again: "Eddie, you're dripping."

He frowned, snapping, "I know. I heard you the first time. What do you want me to do about it? Don't just stand there looking. Clean it up."

I said, "You clean it up. Don't tell me to clean it up." But I walked into the kitchen to get some paper towels anyway.

"Do you know who you're talking to?" he yelled after me. "Do you know who I am? Does he know who I am?"

When I came back with the roll of paper towels, he said, "Do you realize that you work for me? That I'm your boss?"

"That doesn't give you the right to talk to me that way," I said. I wiped up the wet table and then went back into the kitchen.

When I returned this time, Eddie said, "I was out of line. You're right. I'm sorry." He actually looked as if he meant it.

"It's all right," I said.

The curious thing is, after that he stopped treating me like a gofer. No more quarters in the meter. No more sandwich runs. Maybe it was just coincidence. Maybe not.

My relations with Spanky, by contrast, were fast deteriorating. Since he looked up to Monkey as a father figure, I was guessing that Spanky had turned me into the sibling rival. Clearly he couldn't stand the fact that I was receiving less of a hazing than he had. The days of him playing the benevolent tutor were over; we were in competition now, and I was the enemy. Every little bit of seniority he'd garnered he lorded over me as if, in the big picture, it actually added up to something. He reminded me of some of the film crew people I worked with when I was doing P.A. work back in the early eighties. They were all treated like shit, so they took their resentment out on anyone who was lower in the pecking order. It was hard to believe that at thirty-three years of age I could still let myself get sucked into petty head games. But who would have thought that at thirty-three years of age I'd be answering phones in this Lower East Side tenement? Besides, Spanky had his little agenda and no matter how nice I tried to be to him he wasn't going to be deterred. He was officious, bossy, and nasty. One afternoon he even went so far as to rip the phone out of my hand while I was trying to field a question from a customer.

"What the fuck?" I hissed. "What do you think you're doing?"

He waved me off, saying into the phone, "Yes, sir, you'll have to excuse him, he's a new clerk. Can I help you with something today?"

He had pushed me too far. "Do you believe this shit?" I said to the table at large as laughter rang out all around.

"I should kick your ass," I said to Spanky.

Bob said, "I don't know, Pete, he's got sixty-five pounds on you at least."

"Are you kidding me?" I said. "I'll take him apart. I'll kill him."

Everyone continued to laugh, Bob included. "Listen to this guy," he said. "Harvard turns 'em out tough, huh?"

Bob had a way of poking holes in you. He was from the Henny Youngman-Don Rickles school of humor. He'd go after everyone, and his insults did a lot to defuse the tensions of the office. Watching him and Krause tangle provided particular comic relief.

The day before, Bob and I'd been working the day shift together, and Krause staggered into the living room around noon wearing his usual T-shirt and soiled underwear outfit, his hair a seething mess. He said to Bob, "You know, I've been trying to think of the one word that describes you, Bob, and I was looking through the dictionary yesterday and I finally found it. *Oaf.* You're an oaf."

Bob shrugged. "You actually spent time on that, Krause?"

"Don't flatter yourself."

Krause hovered, and Bob looked at me and waved the air in front of him as if to disperse a bad smell. "Krause, you know that square white thing you got in the bathroom? It's called soap. Do us a favor. Use it."

Krause swung his head to the side in an exaggerated way, as if he'd received a devastating blow. "The wit!" he exclaimed. "It's frightening. Don't tax your brain too much, oaf. You might bust it."

The apartment door opened—always a shock when it was unexpected—and Ralph, Krause's best friend, joined our little roast.

"Don't you fucking ring?" Krause yelled at him.

"Why should I? I got the key," Ralph said. He was the guy I called on Saturdays and Sundays to read the pinks to, so in case we got busted we wouldn't lose the action. He was an otherwise pleasant-looking fellow who had about four teeth in his mouth.

"The key's for when I'm not here," Krause said. "When I'm here you should ring."

"Oh," Ralph said. "Well, then, I guess you don't want this."

He'd brought over about five thousand dollars in a brown paper bag, money from Lemon Drop, which he now held in his outstretched hand. Krause grabbed the bag. It turned out that Lemon Drop and Toledo were Krause's players. This was a revelation to me. I watched Krause take the money out of the bag and count it very slowly. I was thinking, This is weird. What if he was actually the boss of the operation? Like Vincent the Chin, the eccentric Mafia chieftain who strolled around Little Italy in his bathrobe.

Ralph said, "Can't you count any faster?"

Krause stopped counting entirely. "Ralph, if you don't shut up I'll knock the rest of those teeth out of your mouth." He turned to me and Bob. "Nine years this guy is paying a grand a month for an apartment. I keep telling him, 'Get a cheaper place and fix your teeth.' He could have bought himself fifty sets of teeth with that money."

"I don't care," Ralph said. "I'm happy with these teeth."

"Lose one more and you won't even be able to call them teeth. You'll be saying, 'I'm happy with this tooth.' "

"What do you care?"

"I don't. But don't you want girls?"

"I have girls."

"Yeah, sure. What girl is gonna kiss that mouth, you snaggle-toothed motherfucker?"

At this point Pat walked in, muttering, "I hate bookmakers," and the whole scene began to take on the dimensions of some bizarre bookie sitcom.

Ralph said, "Next time I have money, I'm not going to fix my teeth. I'm gonna get a place in Vermont with a couple of chickens and a pig."

"Take Bob with you," Krause said. "Then you'll have two chickens, a pig, and an oaf."

49

MY mom and stepdad had returned from the Cape the previous night. We had dinner together, and while my stepdad was watching MacNeil-Lehrer, I hung out in the kitchen with my mom. I'd been planning to keep her in the dark about the bookie stuff, but when she asked me how I was managing, the best I could come up with was some vagueness about a part-time research and clerical gig for a guy downtown.

She looked at me like someone who had seen me in my diapers and knew when I was full of shit. "What are you really doing?"

"What do you mean?"

"Are you working for a bookie?"

I had to laugh. Direct hit, Mom. What could I say? She could read me like a bookie.

We briefly discussed my new job. Aside from a quick frown of disapproval at the outset, she was fairly stoic. I may have been her son, but I was also an adult. Or so the theory went. Mainly, she wanted to know about the risks, which I minimized, and what the long-range plan was.

The long-range plan. That again.

I felt sorry for her. Here she'd done her best, tried her hardest. And look what she had to show for it.

"I think, for the time being, we won't tell your stepfather what you're doing," she said.

Sure. Why add to her feeling of shame?

50

I heard the raised voices from outside Krause's door. "You fuckin' stupid shit! What the hell were you thinkin'?" I turned my key, wondering what the cops would make of Krause's Save the Whales and Greenpeace stickers when they positioned their hydraulic battering ram against the green-painted door.

"Why don't you guys talk a little louder?" I said, entering and seeing Monkey, Spanky, Michael, Bob, Brandi, and Pat gathered around the worktable. "I couldn't quite hear you from the elevator."

They ignored me, continuing in heated voices. It turned out that Monkey had "gotten fucked": the new office he'd rented for us on Canal Street was a room measuring only eight by seventeen feet. "I can't believe you actually put down four dimes on a fucking hallway," Pat grumbled while Spanky tried to map out on a yellow pad various ways for eight people to work at a table or tables in a room that size.

"We'll each have to fit in a space about two feet wide," Michael said.

"So we'll be sure to take showers," Monkey said.

"A hallway," Pat recited. "A fucking hallway."

"Don't come," Monkey said. "If you don't like it, don't come. When I started off, I was working in a boiler room, sitting on top of an over-turned bucket writing bets on matchbook covers."

"What's your point?" Michael asked.

"My point is, the business is going to run out of there. Whether you work there or not, it's up to you. I don't care."

"You're not going to want to work there yourself," Michael said.

"So I won't work there," Monkey said.

Michael shook his head. Monkey was just being Monkey. For him the world was simple. It was this way or it was that way. Adapt or don't adapt. But don't bore me with your reasoning.

His response had been the same when I'd aired my misgivings about renting an apartment to use as the backup office. "You wanna do it, do it. You don't want to do it, don't do it." It was clear that Monkey was more than a boss to some of us. Who could have mistaken Spanky's term of endearment for him—Pop—for anything other than what it was.

I understood. Completely. There was something about Monkey that made you want his approval, made you think he had the answers.

Bob: "You think being a bookie means you wind up going to hell?"

Monkey: "There ain't no heaven and hell. You're a bookmaker. What are you talking goofy shit for? You're here, then you're gone. What else is there?"

Bob: "But what about the guys who fuck you?"

Monkey: "Nothing. They fuck you. That's life."

I'd look at Monkey sometimes and try to imagine what it would be like having him for a father. He might be a criminal, but he radiated such confidence and certainty about how the world worked, not just in his words, but in his manner, in his actions, that it was possible to come to the conclusion that morality really was for suckers.

I thought of my own father, who was always eager to share his theories and ideas about how the world worked but who was nearly hopeless when it came to putting them into practice.

Maybe I wouldn't develop a cement-mixer voice or start wearing tasseled loafers. But was I wrong to think there was something to learn here, something that up till now I hadn't been taught?

51

THERE was an office pool on who would be the first among us to croak. Bernie was the favorite; Eddie was close behind. On Friday a long shot almost came in.

The air conditioner was shorting out, tripping the circuit breaker and shutting down the lights, and Monkey was bitching, saying we definitely had to get a new a.c. Brandi, the Josephine the Plumber of bookmaking, though in her case it was electronics, not plumbing—she went to the Apex Technical School—announced she could fix it. She fiddled around with a Swiss Army knife, gave her a.c. theories, and lo and behold, the thing began to work.

"She's got the magic touch," Monkey said.

Six guys sitting around helpless, ready to say fuck it and buy a new machine, and this blond chick fixes it.

Brandi's glory was short-lived. Ten minutes later the circuit breaker tripped the lights off again.

"What the fuck?" Pat shouted. "What'd you do to the a.c.?" He said it as if it had been working perfectly before and she broke it.

Undaunted, Brandi rolled up her sleeves and took another crack at it. No one was paying much attention to her. All of a sudden there was this loud pop and crackle, the lights blew out again, and a live spark jumped clear across the room toward me, burning a large hole in my yellow legal pad. "Holy shit!" An acrid creosote smell filled the air.

Brandi knelt there in the dimness, stunned, holding her Swiss Army knife, its blade melted. She'd cut into the line while it was still plugged in. Spanky flipped the circuit breakers again, but only some of the lights in the room came back on. She'd blown the whole socket out. There was a long black streak on the wall above it.

"Jesus Christ!" Monkey yelled. "Are you fucking crazy? Cutting into a live wire? You could have killed yourself."

Brandi couldn't speak.

"Are you okay?" I asked her.

Her breathing was jagged and her eyes loopy. The hand with the Swiss Army knife was shaking.

"Brandi?" I waved my hand in front of her. "I think maybe we better take her to a hospital. She doesn't look so good."

"Go get a Hefty bag," Monkey said.

"C'mon, I'm serious. She got a bad shock."

"I didn't realize there was a split in the wire," Brandi said.

"She's talking," Monkey said, "so we know she's not dead."

"We couldn't have been that lucky," Pat muttered.

I continued kneeling by Brandi, but the rest of the guys had already gone back to answering the phones. Couldn't lose that business.

It was clear to me that she was in no condition to continue working, but I couldn't persuade her to go to a hospital. In the end, I helped her downstairs and flagged a cab.

"You're a sweet guy," she said, as I helped her get in.

"Saddam Hussein would be a sweet guy compared to that crew."

"No, you are," she said. She gave my forearm a squeeze.

Back upstairs I was met by grins and snickers.

"My hero," Pat said, making his voice go high.

"We were taking bets on whether you were going home with her," Michael said. "Bob made it two to one."

I flipped them the finger.

"What happened, she turn you down?" Spanky said.

"Fuck you."

"He's just jealous, Petey," Michael said.

I sat down at my phone, shaking my head. Every time I looked up, someone was grinning at me.

"You at least get her to suck your woody in the stairwell?" Spanky asked.

More titters.

It was hopeless. I'd brought it on myself. The worst thing to do was to fight it.

I forced my lips into the shape of a smile.

"Look at him," Pat said. "Smiling like the cat who ate the canary."

"You mean like the cat who ate the pussy," Spanky said.

52

I took Saturday off and went to visit my dad and stepmother in Ossining.

They lived in a little house on Spring Valley Road that was badly in need of upkeep that they couldn't afford. The terracotta-colored linoleum was curling up in the comers of the kitchen, the green-leaf wallpaper had come unglued in the hallways, the furniture was old and covered with dog hair. When my dad saw the way I was looking around, he said it really didn't make that much difference because it had been ages since they'd had company over. My stepmother was out of the house most of the time due to her job. "She works too much," my father complained, implying that she did it not out of necessity but rather to avoid spending time with him. But he too was out of the house now, working again after a couple of years of unemployment. He was in telemarketing, selling jewelry over the phone. At the age of seventy-three, he was not thrilled to be working and certainly not in phone sales, but with a college-age daughter, my half sister, still dependent upon the two of them for financial support, he had no choice.

We took a stroll around the unmowed front lawn before supper, and he stopped at the leafy overgrown edge of the property, turning back to look at the house. He appeared smaller to me than he had a couple of years before. His head seemed shrunken, like a peach that had been sitting in the refrigerator for too long.

I knew that it wouldn't bother him very much to find out how I was making my living. He was different from my mother that way. Still, I hesitated to tell him because the jobs that he and I were doing were not

dissimilar—was it much of a stretch to say I was in telemarketing too?
—and it was painful for me to acknowledge this or to let him know it.

I felt bad that his life hadn't worked out better. It all got so knotted
up, this father-son stuff. All the anger and guilt and regret on both
sides, the inability to change each other. I had for a long time held
him accountable not only for who he was but for who I was.

Yet we were easier with each other now than we had been in years
past. There had been no heart attack or major illness to jolt us awake;
it was just that he was getting old and we both knew it.

Standing there on the lawn, looking at the flaking, deteriorating
house, my father remarked that this wasn't how he'd imagined things,
having to work at this point in his life. Not that hawking jewelry over
the phone was a bad job. He'd had worse. The people at the company
were nice; they seemed to like him. But really, he wished he didn't
have to work at all. He wanted to travel. To go to Europe one more
time.

It hurt me to hear him. To think that, with time precious, he couldn't
do what he wanted. It hurt me because there was no way I could help.
And because—whether I blamed him or not—I was worried about
winding up in the same boat. Unless, of course, I started to make
some money.

That was it finally, wasn't it? Money. If I could make some money,
things would be different. Money as an agent of change or liberation
was a horrible, banal concept, one I had denied nearly every day of
my adult life. Yet I believed it to be true now. I looked at my dad
and I believed it. Maybe that's why in telling him what I was doing,
I emphasized how much potential there was to make money. "You
should see how much these guys rake in, Pop."

I wanted to offer him something. I wanted to say to him, as I wanted
to say to Anna, that I would try to make things better. This was
no tearjerker story. No black-and-white 1940s melodrama. My dad
didn't need some expensive operation; the bank wasn't foreclosing on
the house. He just wanted not to have to work. He wanted to travel.
And I wanted to say, "Yes. Yes. That can happen. That can definitely
happen."

53

WHEN I arrived at the office on Sunday morning, Spanky greeted me so cheerfully I knew something was wrong.

"Where were you yesterday, Pete? We tried reaching you."

"I was out of town," I said. "Why?"

"Your guy went a little crazy."

"My guy?"

"Tool."

I looked over at Bob, who was just getting off the phone. His expression was full of reproof.

"What happened?"

"Listen, motherfucker, is this guy good for his money?"

"Yeah ... I think so. Why?"

"He started going nuts on the late games yesterday, chasing his money. We tried reaching you, but we couldn't."

"Well, what happened? How much did he lose?"

"A lot."

Spanky picked up the Xerox copy of the weekly figures and happily read off the exact amount: "Minus two thousand five sixty-five."

"*What!*"

"You didn't give us a limit for him," Bob said.

"Well, I gave *him* a limit!"

"You gotta tell us. We can't stop him if we don't know what his limit is. What is it anyway?"

"A thousand."

"Oh, boy. You know how to reach this guy?"

"Yeah. I got his work number and his beeper number."

"You might want to start dialing."

I called Tool's work number first and left a message on a machine. Then I tried beeping him. Half an hour later I still hadn't heard back.

Bob said, "If I were you, I'd be getting pretty goddamn nervous right about now."

I was. Every time one of the phones rang, I looked expectantly at whoever picked it up, and I felt the same kind of disappointment when it wasn't Tool that I would have if it had been Anna I'd been hoping to hear from.

Tool finally called ten minutes before the four o'clock games started. "There you are," I said. "I beeped you four times. Didn't you get them?" He said no. It turned out I was beeping him without punching in a number. I didn't know you had to. I said, "You lost a lot of money yesterday. Is that going to be a problem?"

He said, "It shouldn't be."

I said, "It shouldn't be? Or it won't be?"

He said, "It won't be."

He wanted to play one of the four o'clock games. I didn't want him to, but I was afraid of offending him, afraid that if I said no, he might decide that was grounds for not paying what he already owed. I told him he could have a small play.

"What's the Giants game?"

"Four and a half."

"Give me the Giants for a dime."

"I can let you have a nickel, that's all."

He said okay, and I wrote up the slip. We agreed that Wednesday would be our settle-up day. He'd call me in the morning to arrange a meeting place.

After I got off the phone, Bob said, "You feel better, Pete?"

"A little."

"Well, look on the bright side. If the guy does end up paying you, so will we."

54

BY Thursday morning I hadn't heard from Tool. I'd put in about ten calls to him since Wednesday morning, using the beeper number properly now. The guy was going to stiff me.

About the only good thing that had happened was that the Giants had covered the spread in the late game on Sunday, which meant that Tool's balance had been reduced from $2,565 to $2,065. It was actually as if I had bet $500 on the game myself, since I was responsible for the money. If the Giants hadn't covered, I'd have been out $3,065—a thousand-dollar swing that went my way.

Bob wasn't especially sympathetic when I told him about the unreturned phone calls.

"You got okeydoked. The guy okeydoked you."

"You're still going to make me pay?"

"That's the way it works, Pete. You knew that."

"It doesn't encourage a guy to go out and get players if he ends up being penalized like this."

"I told you to check this guy out first before taking his action."

"You told me you'd help check him out."

"Look, we've all gone through this, had guys lay down on us. It's part of the business. Ask anyone here. My advice to you is to find some good players and you'll have this paid back in no time. Meanwhile, you better start working as many shifts as you can." He paused, then growled the inevitable "motherfucker" and laughed.

A couple of days later I received a call from Tool. He immediately launched into some long, involved tale about how he was in the hospital with a stomach problem and that was why he hadn't called earlier. He was convincing enough that I had to fight the urge to offer him sympathy. Even so, I didn't let him get off without giving me his home address. My connection to him felt too tenuous. I wanted more than a telephone number. He told me the hospital would be releasing him in the next couple of days. He would call me then.

Bob said, "That was the guy?"

"Yeah."

"It's a good sign that he called. If he wasn't going to pay, he wouldn't have called."

"I hope you're right." I explained to Bob about him being in the hospital and how that was why he hadn't called earlier.

"Uh-oh. He told you he was in the hospital?"

"Yeah."

A mournful look passed over Bob's face. "I take it back. This guy's definitely okeydoking you."

"What makes you say that?"

"The hospital story. That's the oldest one in the book."

Before Bob could reply, Spanky handed him the phone— Monkey calling. "Guess what?" Bob said into the mouthpiece. "Pete fell for the hospital story."

He held up the receiver so I could hear Monkey's raucous laughter.

"You mean he was just feeding me a bunch of bullshit?" I said to Bob.

"Nah, I don't know. I'm just yanking your chain. Maybe he was really sick."

55

Pat gave me a ride across town after work.

"Have you ever had to cover for a guy who laid down on you?" I asked.

"Sure." He accelerated around a bus, muttering something under his breath.

"Even if the business had already made money off him?"

He nodded. "It's a tough business."

"No shit. I mean, everyone's egging me on to bring in players. Then I do, and look what happens."

"New gamblers are the lifeblood of the business," Pat said. "Bookies are like vampires. We have to keep finding fresh blood or we die. So there's a lot of pressure on everyone. If you show that you're capable of bringing in business, you become very useful to the company. Plus you make a lot of money yourself. If you can't, you become expendable. Why do you think that goofy Spanky eats so much shit, apart from the fact that he's a miserable human being? Because he's not contributing. Michael and Bob bring in money, so they're the little princes. Money equals respect. That's the way life works."

A taxicab pulled up beside us and Pat glared at the driver. "You fuckin' camel jockey. Go back to your sandpile." Pat looked at me, his mouth twisted and sour. "I hate these fucks." He looked back toward the cab. "Learn English, you fuck!" The light changed. Pat stomped on the accelerator.

"The amount of money those idiots Bob and Michael make is staggering. It doesn't matter what they do, they can't seem to help raking in the dough."

"You're making money, too, aren't you?"

"Yeah. I do all right," Pat said. "At least I do something with my money. I mean, I'm putting my son through college. What do Bob and Michael do? They go on cruises, buy expensive toys, and throw hundred-dollar bills around like they're Al Capone."

I knew Pat had a high acid content—his cynicism was almost endearing once you got used to it—but I hadn't realized how much he resented Michael and Bob. I remembered what Bob had told me about going over to Pat's apartment one time, how Pat was always talking about how great his place was, when in fact it was a dump—a stinking, dirty, cigarette-butts-all-over-the-place dump. At one point Bob opened a cupboard in the kitchen, and Pat went nuts, screaming at him not to, lunging across the room at him. Bob couldn't resist—what could Pat be hiding in the cupboard?

It was peanut butter. Nothing but peanut butter. Jar after jar of Skippy. Maybe fifty jars neatly stacked. Pat was all bent out of shape that Bob had seen them. Like now his secret was out.

So Bob tried to play it down, told him it was okay. So he had a lot of peanut butter. So what? But Pat was all wired and kept asking over and over, "What's wrong with a guy liking peanut butter? Nothing strange about that." And Bob was saying, Yeah, nothing strange. He liked the stuff, too. The thing was, he usually bought it one jar at a time.

I asked Bob what he thought Pat did with all that peanut butter. He said, "I picture him sitting there on his stinking couch watching the Spice channel with a jar of Vaseline in one hand and a jar of Skippy in the other."

I was thinking about all this as we were driving across town, when suddenly Pat started telling me how things had come unraveled for him. I remembered that first day I was in the office, him saying that everyone had a story. Like it was prison we had landed in or a mental hospital, that being there meant we had somehow fucked up. Only with Pat, as he told me now, it had happened real quickly. "One minute I was on top of the world," he said. "The next, it was all gone. In one month I lost everything. My wife, my company, my savings. Then I went crazy. I mean, really crazy. You know I'm crazy anyway. But during that period, forget about it. I was so hair-trigger that anything would set me off. I'd walk down the street, somebody'd brush against me by accident, I'd fucking chase 'em for twenty blocks. It was just luck I didn't kill anyone. You begin to understand about all the crazy fucks walking around when it happens to you. I'm not joking. I could have killed somebody."

"Did you try to do anything to yourself?"

"For about five minutes I thought about it. That's when I went to see a shrink."

"Really?" I was shocked. Pat at a shrink's? "What was that like?"

"I didn't kill myself. He gave me a prescription for Prozac, which really is a miracle drug. At least it saved *my* fucking life. I've been on it for two years."

"Did the shrink just give you drugs, or did you also talk?"

"Yeah, we talked. I told him my fantasies about killing people. About killing my ex-wife."

"You still fantasize about it?"

"Now? Now I give a fuck."

A sedan full of homeboys cut us off, and Pat screeched around them, pulling up beside their car at the light. "You fuckin' asshole mother-fuckin' niggers, jigaboo spear-chucking pieces of shit." He shot them the finger.

One of them turned and glared. I said, "Pat, maybe it's not such a good idea to do that."

"Let 'em try something. I'd be happy to shoot their asses off."

"You have a gun?"

"Right here, my friend." He put his hand on the built-in storage box by the emergency brake. "Nine millimeter. Of course those fucks probably have Uzis and sawed-off shotguns." He shrugged. "Fuck 'em anyway. Let 'em try."

56

THE address Tool had given me was 20 Christopher Street. On Tuesday I went to pay him a visit. I'd worked myself up for a big confrontation. There was only one problem: 20 Christopher didn't exist.

Maybe I'd heard him wrong; I actually convinced myself that was possible. I searched the entire block between Bedford and Bleecker, checking out the mailboxes in the vestibules to see if his name was on any of them. There was some construction going on in an open storefront at 26 Christopher. Thinking that perhaps Artie had named the street because he was doing work there, I asked one of the paint-speckled workers if he knew a fellow called Dowd. He didn't.

I grew enraged. The low-life piece of shit had fucked me! He had lied to me like I was some simpleton! I stormed down the street cursing him. Did he know how lucky he was that I couldn't get my hands on him? The motherfucker. Who did he think he was, fucking with me?

When I showed up at the office later, still steaming, Bob said, "Take it easy, Pete. You'll give yourself an ulcer." He thought it was high comedy that I was talking so tough. Pat was in the office, and for the first time he heard me call my player by his real name.

"Artie Dowd?" Pat said. "Are you kidding me?"

"You know the guy?" Bob asked.

"Artie Dowd is the biggest beat artist below Fourteenth Street."

"No shit? You really know this guy, Paddy?" Bob said.

"Sure I know him. Everybody knows him. Jesus, Pete, why didn't you ask me?"

"I asked Bob," I said in a barely audible voice.

"I told you to check him out," Bob said. "You should have asked Paddy."

I nodded. What I couldn't do was explain why I hadn't asked Pat, unless I was prepared to confess to my greed—that I had been worried that Pat would want Artie to be his player. If I admitted to that, I would be opening myself up for particularly venomous ridicule. Keeping my mouth shut, I was just seen as the poor fool who'd been taken. Who knew? I might even be able to work it for a little sympathy.

57

O N Thursday we moved into the new office, "the fuckin' hallway" that Monkey rented for us. It was in an office building on Broadway right below Canal. We were on the twelfth floor behind a frost-paneled door with the name of our company, M & J Telemarketing, stenciled on it. The space was small and narrow, as advertised, divided into a front and a back, but well lit and clean. Working in a real office, as opposed to Krause's dump, made it seem as if we were involved in a more professional and legitimate business concern. There was a new recording system, wall-mounted melamine marker boards on which to post the lines, and new AT&T phones equipped with little domed lights that flashed silently instead of ringing. Without the ringing, the atmosphere didn't have quite the same frenzied edge.

I'd been working lots of shifts. Eleven in one week. Six crisp one-hundred-dollar bills in my pocket. Or rather in Bob's and Michael's pockets. I had to give back most of my pay as soon as I got it.

I was thinking I might actually recover some of my loss, because Monkey had heard about my problems with Artie Dowd, and it turned out that he knew Tony from Sullivan Street, the bookie that Artie had said he sometimes used. Tony was a mob guy in semiretirement who was booking just for something to do. Monkey said he knew Tony from the old days. "I'll have a talk with him," he said. "See what he knows."

"Really?"

"Yeah, I don't know. Maybe he could do somethin'."

"Like what?" I said. "I mean, I'm just curious.. .."

"You want your money, right?"

"Yeah."

"We'll get your money."

I didn't pursue it further. But my imagination ran wild. Artie Dowd hanging by his thumbs in some bologna plant in New Jersey. Or nailed with his own tools to the door of his apartment, wherever it was. And all because he'd been stupid enough to fuck with me, the connected guy from Harvard.

After the session, probably out of some crime-world sense of no-blesse oblige, Bob and Michael invited me along to dinner. They were meeting a couple of their best customers at Smith & Wollensky, a steak house on the East Side.

On the cab ride uptown, I sat pressed between the two of them while they made calls on their cellulars. When we arrived, Bob dropped a ten-spot on the cabbie for the $4.75 fare.

There was a crowd jammed up inside the door by the reservation desk. The bar was a confusion of bodies, smoke, and high-pitched chatter. An older guy in a black leather topcoat shouldered his way over to greet Bob and Michael. His hair was salted with gray, and he had an acne-scarred face that looked as if it had taken a few punches in its day. "They say an hour," he said. His breath was knockdown-sweet with liquor.

"Don't worry," Bob said. "I'll take care of it." He found the maitre d', whose large smile grew even wider when Bob clasped his hand. Michael, meanwhile, introduced me to the graying pug—Nick—and to Gary, a fellow in a shiny caramel-colored leather jacket who looked like he was pissed off about something.

Bob came back and said they were getting something ready for us. A few minutes later we were led upstairs to the landing, to the best table in the house. The maitre d' said, "Anything you need, you just let me know."

We ordered drinks and settled in. Nick and Gary, it turned out, were the agents for one of the office's biggest sheets. Nick owned a bunch of public pay phones in addition to his bookmaking business. Gary, a

Vietnam vet, was Nick's "associate," which I understood to mean that he took care of collections and payoffs.

I had no idea what to talk about with them, so I drank my vodka on the rocks quickly. Everyone else drank fast too, and before long we were all drunk, on our way to belligerent. The conversational range was confined to two topics, sex and gambling, with a brief interlude devoted to food—translation: steak.

"Luger's has the best steaks," Nick said, opening his palms.

"Sparks," said Gary.

"Luger's," Nick said. "End of discussion."

"You're crazy!"

"Fuck you."

"Don't pay attention to them," Bob said to me. "Order the Cajun rib-eye. It's the best." At thirty-two fifty, I believed him.

By the time the lobster-tail starters came, I had three rounds in me and Bob was shouting—we were all shouting—about some bachelor party he'd gone to where, when he walked through the door, "there was this girl standing on a table, and the groom, with everyone watching, was licking her pussy."

"Yecch. That's one thing I will never do," Gary said, his voice carrying over at least four tables. "I will never eat pussy! Ptew. Ptew." He made spitting noises that gained an added touch of realism when a frothy gob landed on the tablecloth.

"You fuckin' pig," Nick said.

"You can call me whatever you like, but I still won't eat pussy."

Bob and I started laughing, but Gary was adamant. His pronouncement provoked a heated discussion of the virtues and hazards of cunnilingus. Nick recited a rhyme: "Smells like fish, quite a dish; smells like perfume, leave the room." Bob philosophized that when you gave, you got. And Gary continued to rail that any guy who went down on a girl was a faggot.

In the midst of this, our plates were cleared and a waiter came by with a humidor of ten-dollar Monte Cristos and a bottle of port. We all partook, chewing the ends off the cigars and spitting them out, lighting up with exaggerated gestures while the waiter filled our glasses with port. The cigar smoke and the booze had my head spinning.

"Here's to all you pansy-assed faggot rug munchers," Gary said, lifting his glass.

"Women know who the real men are," Nick said.

"Fuck you guys," Gary said. "Real men. I'd like to see what you guys do when you're hanging upside down in barbed wire for three hours."

"Oh, Christ, now it starts. The barbed wire."

"Bullshit," Gary said. "Where were you? Were you there? Were you?"

"Gary, please," Nick said. "Not this again. Please not this. Every fuckin' time we have to have this."

"Fuck you, every fucking time. Why don't you just slap me? You're such a real man. C'mon, slap me." Gary waved his hand toward his chin, repeating the challenge.

Before anyone could brace for it, Nick hauled off and whacked Gary with his open hand, knocking him off his chair.

Bob burst out laughing, and his laughter was so immediately contagious that, no matter how appalled I was, I couldn't stop myself from joining in.

Nick bent over to help Gary to his feet, but Gary grabbed him around his knees and brought him down. The two of them rolled around on the floor between tables, grabbing at each other's clothing, grunting and cursing. Diners at nearby tables gaped. Our table only laughed harder.

Bob finally struggled to his feet and tried to break them apart. By the time the maitre d' arrived on the scene with a few of his sturdier busboys, Bob had separated the two men.

The busboys held the two men apart until they'd calmed down, and Bob apologized to the maitre d', saying, "We'll take care of it. Don't worry. Don't worry."

He and Michael peeled off five hundred dollars apiece in hundred-dollar bills and threw them down on the table, and we all walked out of the restaurant, trying to ignore the fact that every single diner in the place was staring at us. Bob kept laughing the whole time, and

Michael said, "This happens every week. Every week these fucking guys get drunk and fight, and we swear we'll never go back there with them."

At home later on, I tried to fall asleep but I couldn't. I was too bloated with rich food and booze. I thought about Anna, wishing she would come and rescue me, knowing she wouldn't. I put my hand on my limp penis, cupping its pliant warmth. Though I had not said one word through all the crude talk of oral sex, I now found myself thinking about the sweetness of a cunt—her cunt—and I moved my tongue against my lips until my breath grew short. But the taste of cigar in my mouth was too strong for my imagination.

Even when I woke up in the morning, hung over after a restless half-sleep, the taste hadn't gone away. I brushed my teeth; I gargled; nothing seemed to work. For most of the day the cigar taste lingered in the back of my throat, and it wasn't until sometime much later on that I realized it was finally gone.

58

THE clocks got set back in the morning, and early darkness was already falling as I came out of the subway on my way to work. It was only five o'clock.

The mood at the office was dark as well. We'd had another bad week. On the bright side, Monkey let me know that he'd talked to a guy who was going to talk to the guy (Tony), and maybe something would get done about my laydown.

Meanwhile, another bookie got busted. One of the Trannys. Pat was talking about it when he came in, at the same time complaining that he'd had to leave thirty thousand dollars under the floor mat of his car, which was parked over on Avenue B near all the junkies and drug dealers. He wound up leaving the office early to take care of it. Bob observed after he left: "You ever notice how Pat always has to leave early right after someone's office went down the night before? He has money under his mat or he gets the flu or something. He always has an excuse."

Eddie didn't come in until the session was half over, but his lateness had nothing to do with the bust. He was just his usual fucked-up self, swearing about the traffic on Canal and a toothache that was killing him. So Monkey gave him a Darvon for the pain. And bang, Eddie was feeling none. Before long he put his head down on the desk, forehead first, and fell asleep. The office was in full swing, but Eddie slept through it all. When he roused himself as we were leaving, there was this big red mark on his forehead from the table.

59

I FINALLY took possession of my apartment. Except for a mattress, which I had delivered by 1-800-mattres, it was bare walls and bare floors. Whatever furniture I owned was still in storage in Chicago.

Furniture, checking account—I'd left things behind in the way of one who was reluctant to sever ties. It was absurd to think of a checking account as my tie to Anna. But I'd kept the account there—banking by mail when I banked at all, since I was a cash kind of guy now—because I felt that it would be too wrenching to close it. "I'm closing out my account"—I didn't like the sound of that.

At the same time I hated the limbo she and I were in. We were talking on the phone a lot, and we were talking well again, but since we were a thousand miles apart and had no clear sense of where our future together lay, or if we should even have one, the effect, once we were off the phone and the connection was broken, was of waiting. And there is a huge difference between waiting for something you know is going to happen and waiting for something you're not at all sure will.

Friends of mine, convinced that I needed to forget about Anna and move on, were fixing me up; I was sure Anna's friends were doing the same for her. In the gambling world they called it hedging—playing two angles at once to lessen your risk of getting hurt.

On a rare night off for me, my pal Renny Sanger took me to the Fashion Week show at Bryant Park. Inside a Noguchiesque tent lit to dramatic perfection, the rich and famous (Ivana Trump, Rachel Hunter, Matt Dillon) watched the beautiful (Naomi Campbell, Linda Evangelista), and when the flashbulbs stopped popping and the music died, Sanger and I went backstage to congratulate the designer, whom he knew. On the way we bumped into someone else he knew, a very attractive woman— but real-looking in comparison to all the unreal-looking models. Her name was Bettina Shaw, and she was covering the shows for the *New York Observer*. Before we had exchanged more than a few dozen words, she asked me in her slightly British accent if I would mind being her date for the evening, as she had to go to some awful party afterward and needed "protection."

I thought maybe she was asking me because I was a friend of Sanger's, or perhaps because I looked harmless and therefore safe. "What makes you think I'll protect you?" I said.

Her lips formed a subtle smile. "Maybe you won't."

Before long—it seemed as sudden as it was magically improbable—I was on my way uptown to a pricey Fifth Avenue address with her. Bettina Shaw was one of those beautiful young women you meet only in New York, Paris, or London: cosmopolitan, whip-smart, and amazingly self-possessed. I quickly learned that she was the daughter of a countess and a rich industrialist, had been refined at various European boarding schools, and though only twenty-eight had already been married and divorced twice. For her part, she just as quickly discovered that I was a writer *and* a criminal, though she delightedly refused to believe the latter claim.

The party was a study in moneyed stuffiness and chinlessness, the hostess a friend of Bettina's from boarding school. "Don't worry. We don't have to stay long," she whispered, squeezing my hand.

I went to get drinks and lost her for a while. The large, opulent apartment was crowded with people in elegant evening wear. To my amazement, no one looked at me as if I didn't belong, and I had a sudden palpable sense of the range of my life—its stretch from a smoky closet on Canal Street to this grand marble-floored palace on Fifth Avenue. At the bar, a high-strung southern deb in a slinky metallic green backless dress started cross-examining me. Where had she seen me? Cortina? Or was it Aspen?

When Bettina caught my eye from across the room I felt a charge of warmth; it was a feeling I hadn't had with anyone except Anna in a very long time. The nosy, nervous girl with the naked back, whose name was Teddy—short for Theodora—said as Bettina joined us, "This man is very mysterious. Has he told you anything about what he does?"

Bettina looked at me before answering. "He's some kind of criminal," she said, "but I haven't found out what kind yet. I only picked him up at Valentina's show an hour ago."

"Oh, you're as bad as he is," Teddy huffed, waving her arms. Bettina and I left a little before ten-thirty and cabbed downtown to a restaurant she described as "nice and cheap," although in fact it fit only the first part of that description. I had no idea what had possessed her to pick me out of that crowd in the fashion tent, but I realized she'd made certain assumptions that I would not be able to satisfy. At one time in my life I might have tried to live up to her mistaken perception of me. Now I simply made the decision to be real. Up till that point, our conversation had been ironic and flirtatious, and though we disagreed about nearly everything—food, books, writing, art, men and women —I think she found me charming anyway. Sitting across from her over dinner, however, I changed the tone. Later I would regret it, for it occurred to me that underneath the banter, we were both just lonely and horny—and if I'd stayed on track, kept playing my cards right, I could have gotten laid. Down deep, though, I knew I wouldn't have been satisfied to leave it at that. I'd have ended up wanting more from her than she was willing to give. So I let the confessor take over, the one who so freely opened his veins and bled. By the time I finished, the sexual tension was gone. She saw that I was no more than a lost little puppy at an age when I should have been a dog.

I walked her home, both of us a bit boozy, and when we reached her door, I said something about seeing her again. Her answer was a kiss on the cheek, polite but noncommittal.

The next morning I called her—a call that she never returned. I'm not sure what I expected. When I told Bob, he said, "Maybe on the first date it's not such a good idea to talk about you and Anna *and* being a bookie."

"You think it's a bit much?"

"It's just a suggestion."

60

ANNA, with her radar, had called around the time Bettina and I were on our second bottle of wine. She'd left a message that she'd had a sexy dream about me.

"So where were you?" she said in a joking voice when I rang her back. It was near 1:00 a.m. her time. "You sound drunk. Were you out on a date?"

I laughed a little too hard. "No, it wasn't a date. I was with my friend Renny Sanger."

"I don't know her."

"It's a him."

"I don't know him. But it's fine, you know, if you were with a girl. It doesn't matter."

"Anna . . ."

"It doesn't. I don't care."

"Anna, I wasn't with a girl."

She was silent. I imagined her rubbing her palm on her knee rapidly, the way she did when she was anxious. "Oh, Pete," she said, her voice near whiny. "I don't know. I think this is too hard. I just don't see how we can keep this going."

"I thought we were going to be relaxed about it."

"I know that's what we said. But things are too unclear. It's too unsettling. Doesn't it fuck you up as much as it fucks me up?"

"How do we make things clear, then?"

"I don't know. I really don't."

My turn to be silent.

"Say something to me. Say something."

What could I say? That she was right? "How about if we make a plan to see each other?"

"When?"

"Thanksgiving?"

"My sister is coming. And maybe one of my brothers. They're going to be staying with me."

"And Christmas?"

"I haven't decided. I might take Nathaniel to Pittsburgh to see my folks, in which case that wouldn't work either."

"Uh-huh."

"Anyway Christmas is so far away."

"Not that far."

"I get lonely, Pete."

"I know." I slumped forward and closed my eyes. My head was starting to hurt, and I still didn't have anything I could offer her, any realistic proposal to make that would shed light on the future. "I'd really like to see you, Anna. I don't know what else I can tell you. If you can figure out a time . .."

"Can I think about it?"

"Of course."

"I just want to be sure we both know what we're doing."

"Well, I was kind of hoping it would be in this lifetime."

A moment passed.

"That was a joke," I said.

"I know. I just don't think it was very funny."

"Sweetie, it was a joke. We don't need to get into anything over it."

"I'm not getting into anything. I feel like you're accusing me."

"I'm not accusing. Can we just. .. Let's just drop it."

"All right, we'll drop it."

"Anna ..."

"What? I know I'm being a bitch."

"No, that's not what I was going to say. Anyway, it's okay." I tried laughing to bring back the earlier lightness. "Really. Everything's cool. Just think about a time, all right?"

"Okay."

"And tell me about your dream," I said. "Did you really have a sexy dream about me? I want to hear it."

61

O N Monday night after work, discouraged by Monkey's lack of progress in collecting my debt, I went back to the basement of Our Lady of Pompeii over on Carmine Street. I didn't expect to see Artie Dowd there, but through the dim haze of smoke drifting up toward the high ceiling, I looked for him anyway. The haggard, resigned faces, rumpled clothes, and stale stench of loneliness struck me now in a way it hadn't on my previous visit; I sidled over to the cashier's table where two women in cheap flowery blouses were making stacks of tens and twenties out of singles.

"I'm looking for a guy named Artie Dowd," I said.

The two women stared at me. One of them closed the gray metal box in front of her, which was stuffed with cash. The other stopped counting her handful of dollars.

"Who are you?" the one who had closed the box asked. She had a cigarette dangling from her lip and a voice that made me picture her vocal cords being rubbed over a cheese grater.

"I played here a couple of weeks ago, Artie was sitting next to me. He gave me his phone number, but I lost it. I was—"

"Jimmy!" she shouted, so loud it startled me. I looked around, expecting to see some bruiser, and readied myself for flight. A skinny old guy in an untucked blue shirt that was about two sizes too big for him got up from one of the tables and came over. He was wearing thick black glasses that magnified his eyes to about twice their real size.

"This guy's looking for Artie Dowd," she croaked. "You know where to find him?"

The old guy's forehead wrinkled.

"I played poker here with him," I said. "He was gonna help me out with some carpentry I need done."

The old guy fingered his glasses. "I ain't seen him in here lately. Maybe his grandmother could tell you where to find him."

"His grandmother?"

"Muriel Percante. She's got a house on Leroy Street. I don't know the number. Maybe you could look it up."

I thanked him and left, taking the stairs to the street two at a time, thinking, Fuck this bookie shit, I should be a detective.

Back home, I called Information and got the number and address, and the next day before work I dropped by 16 Leroy Street. It was an old brownstone on the block between Bleecker and Bedford. The outer door to the vestibule was unlocked, and inside, to my surprise, one of the frosted-pane double inner doors was cracked open a few inches. I put my hand on the white porcelain knob and opened it enough to peer into a dimly lit hall with a staircase straight ahead to the right. The door to the left of the stairs was partly open and I could hear scraping sounds from within.

"Hello?" I called. "Anyone here?" I approached the door. A drop cloth was bunched up at the bottom of the doorway. I knocked and called out again. Nothing. I stepped over the tuck of the drop cloth and saw a guy on a ladder near the back window scraping sheets of flaking paint off the ceiling. He had a red handkerchief tied on his head, and his back was to me. I inched farther into the room.

"Hey!" I said, rapping on the open door.

He twisted around on the ladder.

"Is Mrs. Percante in?" I asked.

He shook his head.

"Do you know when she'll be back?"

"Nope." He brushed some paint flakes off the top of the ladder. "Her grandson's upstairs. He might know."

"Her grandson?" I tried not to show my shock and excitement.

"You want me to get him?"

"No, no. That's all right. I can get him." Before he had a chance to say anything else, I ducked back out the door and went up the thread-bare-carpeted stairs to the second-floor landing. At the end of the hall, I opened a door without knocking.

The room was musty, the tattered royal purple curtains drawn. It was the room of an old person, cluttered, slightly in disrepair. Artie was stretched out on a four-poster bed in his paint-streaked work clothes, his eyes closed. I creaked forward across the floor, but he didn't stir.

I watched his steady breathing for a moment, trying to figure out what to do. I hadn't expected to find him this easily.

"Artie," I growled, doing my best tough-guy imitation.

He opened his eyes almost immediately and, without moving, looked right at me. "Hey," he said. "Pete. How are ya, Pete?"

His lack of fear was unnerving. How could he be so calm? If I woke up and saw a guy I owed money to standing in front of me, I'd at least flinch. He yawned and stretched, slowly working himself up to a sitting position, one side of his chubby face slightly creased.

"So how come you haven't gotten in touch with me?" I demanded.

He smiled.

"Don't fucking smile," I said. "It's not funny. You owe me a lot of money."

"I've been trying to get in touch with you."

"Bullshit."

"I have. What did you guys do, change your phone number?"

In fact, we had. But I wasn't going to let him off that easy. "Look, you could have called me," I said. "You have my home phone number."

"I tried. I swear to God I tried calling you."

"Fine, you tried," I said. "That's a nice story."

"I did."

"I want my money."

"Don't worry. I'm going to get you your money."

"When?"

"I'll get it."

"I want to know when."

He stuck a finger in his ear, then removed it and examined the tip. "You gotta give me a couple of days."

"You've had a couple of weeks, Artie."

"This hospital thing really set me back. You know how much they get you for in there?"

I swallowed, seeing the impossibility, the absurdity of this encounter. He was so practiced in lying that he was better at it than he was at telling the truth. All my fantasies of this confrontation, the murderous rage I had worked up—who was I kidding? For all the fear I inspired, I might as well have been standing there in a tutu and a mortarboard. I would walk out with nothing. And the minute I did, the whole stupid game would start over again, me on a wild-goose chase, playing punch-button with a phone.

"I'm not leaving without getting something from you," I said.

"Pete, I'd give you something if I had anything on me. I swear it."

"You have no money on you at all?"

"I think I got ten bucks. You want that?"

"No." My disgust was audible.

I didn't want him to feel he was doing anything even remotely honorable. When he removed a crumpled bill from a pocket of his painter's pants and held it out, I just looked at it. "By Friday I can give you some more," he said. "I promise."

"I want to know where I can get in touch with you," I said. "And I don't want any phony addresses this time, like that one you gave me on Christopher Street."

"That wasn't a phony address."

"Artie, why do you have to lie? I really hate that."

"I'm not lying."

"I went over to that place. Don't tell me you're not lying."

"You went to twenty-two Christopher?"

I looked at him with near awe, the baldness of his lie and my utter helplessness to do anything about it apparent to both of us. "You told me it was twenty."

"Twenty? No, it's twenty-two. You must have misunderstood."

"I looked at the mailboxes in twenty-two. Your name wasn't there, either."

"It's my girlfriend's place. I told you that. It's her name on the buzzer."

I don't know why—I knew it would only elicit another lie— but I asked him what his girlfriend's name was.

"Her last name's Martin."

I tried to remember if I had seen it. "If I go over there now, I'll see her name on the buzzer?"

"Yeah. Absolutely."

"And that's where you're living?"

"That's where I'm living."

I was worn down, beaten. "Okay," I said. I took the crumpled ten from his hand and though I was tempted to do something dramatic, like grind it into the carpet or stuff it down his throat, I slipped it in my pocket and headed quietly for the door.

That was the last I ever saw of Artie Dowd or his money.

62

I had paid off about eight hundred dollars of the two thousand I owed Bob and Michael, but with rent to pay now, finances had gotten trickier. One day when I received my weekly pay envelope, Bob stuck out his hand and said, "How much do I get back?"

"I can't this week."

"You're joking, right?"

I waited until the session ended, then told him I'd like to sit down and talk things over with him.

"I hope this isn't going to be some sob story," he said. "You promise me, before I buy you a fucking drink, that I'm not going to have to listen to some bullshit."

I grinned a little sheepishly and said nothing.

We went to an Irish bar on Canal, one of those places where half the customers' heads were lying on the bar next to their whiskey. We took a booth with big rips in the seats and stuffing coming out. A rippled shamrock decal was pasted on the tired mirror that ran along the wall.

"So?" he said.

"If I can't figure out a way to make some more money, it's going to be a while until I can get this paid off."

"Pete . . ." He shook his head, deeply disappointed in me. "What's a while? Never mind. Look, all you have to do is get a couple of good players. What have I been saying to you? Paddy's averaging a few dimes a week for what, eight, nine players?"

"I'm trying."

"You must know *some* guys with money who like to bet."

"I know some pretty wealthy guys. But it's not that easy for me to go around advertising what I do. You know, it kind of compromises my credibility if I—"

"I get it, I get it. It's embarrassing for you. Why don't you just face it, Pete? You're a bookie. That's what you are now."

The words stung.

"All right, all right," he said, "it's nothing to cry about. Listen to me, Mr. Smart Guy, I got a proposition for you. You know all about the other kind of books, right?"

"What are you talking about?"

"Books, literature."

"Yeah?" I had no idea where he was going.

"I want you to teach me."

"Teach you what?" My mouth stayed open in the dumb shape of a question.

"You know," he said. "About books."

"Jesus, Bob, you're a college graduate. You graduated from one of the best goddamn schools in the country. Didn't you read any books while you were there?"

"Sure, math books, science books."

"English is a required course. You must have done freshman comp. That course usually has a decent reading list."

"I hired people."

I laughed. "And you want to learn now? Why?"

"Because I don't feel like being an idiot all my life."

I practically sprayed beer in his face.

"What's so funny?" he said.

"No, I—"

"Listen to me, motherfucker, I'm doing you a favor here. You teach me about books, I'll let you off the rest of what you owe."

"You're kidding."

"Do you want to do it or not?"

We cabbed straight up to the B. Dalton on the corner of Sixth Avenue and Eighth Street. I led him to the Penguin Classics section.

"We start with the Russians," I said.

"Okay."

I suggested that we begin with something he could relate to that was also short: Dostoyevsky's *The Gambler.* He wasn't having it. "I want the real stuff," he said. "Don't pander to me."

"Bob, I don't think Dostoyevsky is exactly pandering." We settled on Tolstoy's *Anna Karenina,* which somehow I had never gotten around to reading. Bob paid for the two copies.

"A hundred pages by a week from now?" I said. "Do you think you can handle that?"

"Let's make it a hundred and fifty. I don't want to spend forever on one book. What's the one after this going to be?"

"We haven't even cracked the cover on this one yet."

"I get impatient," he said. "Once I start doing a thing, I like to do it."

That reminded me of something. "Whatever happened to your idea about taking guitar lessons?"

"I'm taking 'em. You didn't know that? I go three times a week. In another month my teacher has me scheduled to play in a little recital."

"I had no idea."

"Wait'll I sink my teeth into this guy Dolstoy," he said. "Then you'll be really impressed."

"It's Tolstoy," I said. "Tolstoy and Dostoyevsky."

"See? This is great. I'm learning shit already."

63

For weeks I had been racking my brains trying to think of people I knew who bet.

Then one day I got a call from Charlie Rosenbloom and suddenly I had not just a player but a big player, a franchise player. It turned out Charlie's brother knew a fabulously wealthy movie producer who liked to bet football. He was sending him my way.

There turned out to be more to it than that. This wasn't just a peace offering from Rosenbloom (the two of us hadn't talked since he'd left town); it was a bribe. Charlie and his brother had started working on a script together. Our script.

I suppose if I had thought there was much chance of my going back to the script, I'd have been more upset. As it was, I put on a bit of a scene, but it was mostly knee-jerk, obligatory. After I had conveyed my displeasure, I told Charlie that I'd better at least get a story credit, for chrissake, and he said sure, absolutely. But the bribe had mollified me in a more substantial way. Even though I saw the irony of the trade-off, I was happy to be getting something tangible and of more immediate value to me.

At the office I proudly announced that I had a new player—Mogul for Topsider.

Pat rolled his eyes. "Sounds like another laydown to me. How much you still owe on that other one?"

Spanky snickered. I looked at Bob, who gave a near-imperceptible shake of his head. We had our little arrangement, which even Michael didn't know about.

"That guy still never paid you, huh?" Monkey said.

"No." I had made up my mind not to say anything about my encounter with Artie Dowd.

"Petey, you check this new player out?" Michael asked.

"He's a Hollywood bigwig, a very respectable citizen," I said.

"Those are the worst kind," Pat said. "This is a guaranteed laydown."

"You guys can laugh now," I said. "But you'll see."

"Yeah, we'll see," Pat said. "You'll be handing over your pay envelopes for the next five months. That's what we'll see."

When Mogul called a couple of days later, it was Eddie who picked up the phone. "Does anybody know Mogul?" he asked.

Spanky cracked up. Eddie was lucky to know what day it was. I grabbed the phone from him. "Hi, how ya doin'?" I said, trying not to betray too much eagerness. I had talked to Mogul from home several days before, and we had agreed on his limits. When he made several two-dime bets, I realized I was nervous as shit. This guy was playing big. What was I going to do if he didn't pay?

The others took a look at the ticket after I tossed it into the box and either raised their eyebrows or whistled.

"You guys really suck, you know that?"

"Hey, we just don't want you to get hurt, Petey."

On one level that might have been true, but on another I felt that they would have liked nothing better.

Mogul lost six thousand dollars his first week, which was a thousand over our agreed-upon settle-up number. But Monday afternoon he called to check on his figure just as he was supposed to, and when I told him what it was, he said, "No problem. Just tell me how you want it settled." Steak Knife had a guy in L.A., I told him. He would get in touch. They could make arrangements.

Three days later—three admittedly sweaty days for me— Mogul met with Steak Knife's man and made good on what he owed.

"Breathing a little easier?" Paddy asked.

"Nah." I was dismissive. "I mean what's six grand to him?"

"Well, it looks like you got yourself a franchise player."

For me, that six thousand dollars translated to six hundred. That was my ten percent. But the first time I got Bob alone, in the same low-life Irish bar on Canal—we'd returned there to launch our literature class —I renegotiated.

"I got a great player, but I'm taking too big a risk," I told him.

I'd expected a little bit of an argument, but Bob didn't even blink. "Okay. Twenty percent. I'll give you twenty."

"Really?"

"Listen, motherfucker, is that enough or not? Because I don't want to talk about money. I want to talk about *Anna Karenina.*"

"Twenty percent is fine," I said.

"Good. I've read up to page two hundred, and we've got a lot of ground to cover." He whipped his copy of *Anna Karenina* out of his ever-present gym bag.

I had only made it to page 130, but I lied and told him I'd read through 150, as we'd agreed upon.

"I guess you're liking it," I said, "if you read that much."

"Well, as you know, I don't have much to compare it to. But I think this Tolstoy guy is no ordinary genius."

I laughed. "Yeah, that's a good way of putting it."

"I mean, the people .. . they have contradictions, complicated feelings. . . . They're so *real.*"

I agreed, and the two of us rhapsodized about Levin and Oblonsky and Anna and Vronsky and the ways in which Tolstoy had managed to make them so real. I hadn't been at all sure of the sincerity of Bob's desire to learn about books, and so his gruff enthusiasm was both surprising and oddly moving to me. Also, his response to my own rather pedestrian interpretations of the text made me feel as if I were Moses on the Mount. When I talked about Anna's transformation after meeting Vronsky, the way she sees everything in her world anew, how she suddenly notices the boxy size of her husband's ears, which had never bothered her before, Bob got so excited—"This is *great!* This is *great!*" —that I began to wonder if I had missed my true calling.

"Listen," he said, when he was paying the check, "don't tell the other guys about this. That we're doing this. I want it to be our secret."

"Why? You worried about what they'll say?"
"I don't know. It's not that. Let's just let it be our thing."
"Our thing? You know what 'our thing' is?"
Bob looked at me a moment, then nodded, realizing. "Cosa Nostra?"
"Exactly."

64

EVERYONE in the office seemed to be in general agreement that Eddie was a near incompetent as a bookmaker. It didn't help that he was almost always stoned or coked up. But because he was the big boss's brother, he'd been put in charge of charting, making line changes, and betting out. This irked the shit out of Monkey, who had his own cockeyed theories about moving the line and betting out and wasn't inclined to hold back his opinion.

Friday night we had taken the worst of it on a few games, and Monkey was ragging Eddie nonstop, rasping at him over and over, "What are you doing? You don't know what the fuck you're doing," until Eddie finally cracked and said, "Okay, you wanna do it? You want to sit here?"

"Yeah," Monkey said. "I can't do any worse than you're doing."

"Well, forget it," Eddie sputtered. "Christ! You think I don't know what I'm doing? I know what I'm doing! I've been doing this for fifteen years!"

"So how come you're doing it wrong?"

"I am not doing it wrong. What are you talking about?" Eddie picked up some of the bets he'd just made and waved them in front of Monkey's face. "You don't want these?"

"No, we don't need 'em," Monkey said. "What do we need 'em for?"

"Fine, you want me to give 'em away?"

"Yeah, give 'em away. Give 'em away to your sister. Maybe she'll know what to do with 'em."

Eddie started spitting he got so mad. His words grew unintelligible, and flecks of saliva collected on his mustache. His head and body trembled. Suddenly he picked up his briefcase and threw it across the room. It hit the wall, papers flying everywhere.

The next day an emergency meeting of the partners was convened. At Bob's suggestion and as a compromise, Bernie was put in charge of charting and line changes. Apparently Bob had been pushing for the move for weeks. "Bernie is sharp as shit," he explained to me afterward. "You wouldn't know it to look at him now, but he was very rich at one time."

Bernie had worked on Wall Street in his younger days, made over three million dollars, then gotten involved in an illegal bearer-bond scheme that had landed him in Allenwood for a year and left him bankrupt. Afterward, unable to work on the Street again, he'd gone to work for a bookie and within two years had taken over the guy's business. As on Wall Street, he made a ton of dough very quickly, then self-destructed again, gambling it all away. It was during this time that he'd accumulated a hefty debt to his boyhood pal Monkey, a debt that he was still paying off by working at the office.

I asked Bernie once about his compulsive gambling, whether he'd ever tried to stop.

"Yeah, my wife made me go to GA meetings for a while. But they didn't do any good."

"No?"

"I don't really want to stop; it's the only thing I like doing."

"What about your wife?"

He shrugged. "One time we were lying in bed, watching Monday Night Football, she says, 'Bernie, you got a bet on this game?' I say, 'No. I swear on my life, Lucille, I'm not betting anymore.' She looks me right in the eye and says, 'Are you absolutely sure?' I tell her, 'I'm sure. I'm sure. May God strike me down right here.' Anyway, end of the game, the other side is driving the length of the field, I'm going crazy, but I can't say anything. Lucille's right there in bed next to me. I'm gripping the headboard tighter and tighter. Finally the other side's got the ball on the one-yard line, there are a couple of seconds left in the game, and Lucille says to me, 'Look at your hands. Jesus, look at your hands!' So I look at my hands, and my goddamn fingers are bleeding! I'm squeezing the headboard so tight my fingers are bleeding!"

"So she knew?"

"Of course she knew," he said.

"What did you do?"

"What do you think I did? I told her to wait until the game was over and then we'd talk about it."

"What about now?"

"Now she doesn't give a shit. The title to the house is in her name, she's got her own bank account, our finances are completely separated. And I gotta pay her rent every month or she'll kick me out. Basically, I'm a boarder in my own home."

65

THE week before Thanksgiving I made a small killing. Mogul lost $19,000, and thanks to my new deal with Bob, I pocketed $3,800 of it. Thirty-eight one-hundred-dollar bills. I hid a thousand of it in a Charlie Parker CD case and another thousand in a Rickie Lee Jones CD, and with the rest, I bought a couch from Conran's, a dhurrie rug, a beautiful flea market antique wood desk, and a TV and VCR. I bought some Christmas presents for my family and a hundred dollars' worth of groceries, and I still had enough cash left to take an old girlfriend to dinner at Bouley. Like the wise guys always said, money won was twice as sweet as money earned.

It was really amazing, at least in the short run, how a little windfall could cheer a guy up. Long-term I knew better than that. Or at least I should have. But it was fun throwing money around. I told Anna about it, omitting the detail of the dinner at Bouley, and we had such a light and giddy conversation that she called me again the next day to tell me that she thought she'd be able to leave her folks' house in Pittsburgh after Christmas and come spend a few days with me.

Now that I was making some dough—in addition to Mogul, I had gotten a couple of dollar players, also through friends—I expected that the office hazing would slack off a little. Instead, I got kidded more than ever. The only difference was that after five months I had learned to dish it back. Still, some of what I had to listen to was pretty rude. Pat decided one day, watching me sit there daydreaming while my phone was ringing unanswered, that I'd been lying: I hadn't gone to Harvard University, I'd gone to *Howard*. Pretty soon it became, as did most things that initially drew a laugh, a running joke. Monkey, who hadn't been there at the joke's inception and didn't quite know what to make of it, asked if it was true.

"Look at him," Pat said, pointing at me. "Don't you see the kinky hair? He's black. He's a nigger. It's Howard, not Harvard. *Howard.*"

Monkey practically died. I knew that to respond would only serve to incite them. But the joke got old fast. Finally I said, "You guys are like little children. You can say the same dumb thing over and over and never get sick of it."

Monkey and Pat both nodded soberly. Monkey said, "You're right," waited about ten seconds, and added, "*Howard.*" And the two of them laughed so hard they fell off their chairs.

A couple of days later I found myself alone with Pat before a session and I decided to tackle the "nigger" thing directly. I didn't mind them calling me Howard, I told him, but I did mind when it began to devolve to "nigger." That was crossing the line.

"Hey, you know we're only joking, Petey," Pat said.

"I know, and it's not that I can't take a joke. It's just that I find that particular word offensive."

Pat got all bristly. "All right, I'll stop."

"I know it's not personal or anything. It's just the word."

Pat looked like a steam valve in a cartoon, swelling and contracting at an alarming rate. "Listen, I'm not going to stop using the word. So don't ask me to. I'm a fucking racist and I don't like niggers. So don't even think about asking me to stop using the word."

I kept my mouth shut. Pat was so tightly wound that all I could think was, Thank God this guy doesn't work in the post office.

He actually did stop using the n-word for a few days, but there was plenty of other stuff for him to give me a hard time about. Like Brandi.

Ever since I'd put her in a cab after the air conditioner incident, Brandi had decided I was the last good guy on earth. Despite the shit I had to take for it, I kind of liked the role. Between shifts on the long football weekends, Brandi and I would sometimes go over to the flea market on Broadway right above Canal. I'd scope out junk for my still mostly unfurnished crib, and she'd advise me. Brandi was the only clerk in the office who consistently fucked up writing tickets—probably because she was talking too much—but when it came to dickering with the vendors, she was all nails. More than once in mid-negotiation she grabbed my elbow and yanked me away. And more than once this tactic worked to lower the price on an item by fifteen or twenty percent.

During one of our little Chinatown-SoHo expeditions, Brandi told me all about her no-good boyfriend, Jasper, the subway motorman. I thought of them as the motorman and the motormouth. Apparently Jasper was a stud in bed and was extremely nice to the kid. But she found out he'd cheated on her with some tramp who worked at Cameo Nails & Tanning, and Brandi wasn't going to put up with that kind of shit. "He knows it, too," she said. "I don't let people treat me bad, Pete. I can forgive somebody once—everybody makes mistakes—but he better not do it again."

Brandi didn't feel too charitable toward Anna based on what I'd told her. "You're a really good guy, Pete," she said. "And there ain't many of them around. Not nearly enough. If she doesn't figure that out, it'll be her loss. That's how I feel about Jasper and me. He doesn't figure it out, it'll be his loss." She told me how the kid, Rollie, had helped open her eyes that way. "He knows he doesn't have long, so he appreciates things. In a way, I'm more excited about this trip to Australia than he is. Just to get to see *him* enjoy it. . ."

On our way back to the office the Saturday after Thanksgiving, the attendant at the lot where Brandi parked her car yelled for her to come over. Apparently her boyfriend had left a note in the car.

Brandi frowned. "Jasper came by and left me a note? That's weird."

I could see she was thinking, Was it about Rollie? I walked with her to her car, a blood-red Camaro with a lot of miles on it. There was a folded-up piece of yellow legal paper on the front seat. "I don't understand why Jasper would leave me a note," she muttered. She opened the driver's door and I watched her unfold the note and read it, her face reflecting first consternation, then fear.

"Jesus, what is it? What does he say?"

She swallowed, at a rare loss for words, and handed the note to me. It was written in blocky capital letters with a ballpoint pen.

WANT ONE HUNDRED LARGE BY TOMORROW OR WE MAKE BIG TROUBLE FOR YOU—COST YOU TWENTY TO FIFTY IN FINES AND LEGAL FEES AND YOU LOSE BROADWAY AND 156 ST. MARKS AND WE GO TO YOUR HOUSES AND GET WHATEVER WE CAN. IF YOU GO SOMEPLACE ELSE WE WILL FIND YOU IN MINUTES. WE DON'T STOP UNTIL YOU ARE OUT OF BUSINESS. IF YOU FORCE US TO COME WE USE CALLER ID SETUP, TAKE ALL BETS, GET EVERY CUSTOMER'S NUMBER. HAVE BEEN TAPING ALL CALLS, HAVE VOICES AND EVEN YOUR GARBAGE FOR WRITING SAMPLES. . . .

I stopped reading and looked up, suddenly feeling very exposed, a throbbing fear in my chest squeezing the breath out of me. There were cars surrounding us, some of them on top of hydraulic lifts, other cars parked underneath. People in winter jackets strolled by on the sidewalk. Yellow cabs careened down Broadway. A shiver went through me. I continued reading.

IF WE HAVE TO COME, WE COME WITH TOYS SO OTHER CHARGES WILL BE MADE AGAINST YOU. ANY CLERKS NOT WORKING WHEN WE COME, WE GO TO THEIR HOUSES AND GET THEM OR WAIT UNTIL THEY COME HOME. WE IMPOUND CARS AND USE RICO. WE THINK YOU GUYS AND GALS HAVE BEEN DOING PRETTY GOOD. SO WE WANT OUR END. HAVE ALL CLERKS—ALL OF THEM. NO EXCEPTION —ON WHITE AND BROADWAY TOMORROW MORNING AT 11 A.M. HAVE THE PACKAGE READY. IF YOU DON'T HAVE THE FULL PACKAGE, DON'T BOTHER SHOWING UP. WHEN THE PHONE RINGS, FOLLOW INSTRUCTIONS. IF SOMEBODY'S MISSING, NO DEAL. WE COME. HAVE THE CAMARO AND CONCORD READY TO ROLL—LUCKY ANDY.

"This is unbelievable," I said. "Fuck." Monkey's Concord wasn't even in the lot. He liked to park it out on the street.

"Do me a favor?" Brandi said.

"What?"

"Can we not tell them we found it in my car?"

"Why?"

"I don't know, I just have a feeling they're not going to like it."

"I know they're not going to like it, but it's not like it's you. So what if it was in your car?"

She shrugged, not very happily.

The parking attendant was inside his booth, his hands hugging a steaming blue-and-white paper cup of coffee.

"Hey," I said.

"Whassup?"

"The guy who left this note—what did he look like?"

The attendant was a light-skinned black guy with amber eyes. He turned them on Brandi, then back on me. "What? Did I do something wrong?"

"No, nothing. Just wondering if he said anything to you and what he looked like."

"Man just asked me if he could leave a note in his girlfriend's car."

"Did he say her name?"

"No, he just, you know, said the blond girl who drives the Camaro."

"What did he look like?"

"Big white guy. Brown hair, mustache. I never seen him before."

I looked at Brandi. She shook her head: it wasn't Jasper.

66

Monkey read the letter a second time. "This is no fuckin' good," he said. "We gotta get out of here."

"Right now?" Bob asked.

"I don't know about you," Pat said. "But *I'm* getting the fuck out of here before anyone has time to ask any more dumb questions." Pat gathered up his pink betting slips and other paraphernalia and stuck them all in his black shoulder bag. "I'll call Steak Knife when I get home," he said. "Color me gone."

The door slammed behind him.

"Big fuckin' pussy." Bernie puffed a cigarette like it contained pure oxygen.

"He's right, though," Monkey said. "We gotta get the fuck out of here."

They had already made me and Brandi run through the whole scenario twice: how the lot attendant had called us over, his subsequent description of the guy, whether we'd seen anyone hanging around, and so on.

"Who do you think it is?" I asked.

"Whoever it is, they know shit they shouldn't know," Bernie said.

"It sounds to me like cops," I said.

"Crooked cops," Michael said.

"The language is weird, though," I said. "Like it was written by somebody Chinese."

"Or somebody trying to sound Chinese."

"The guy said to the attendant, 'I want to leave a note for my girl-friend'?" Bob said.

"That's what he said."

"All right," Monkey said. "Enough talking. Let's get the fuck out of here. We'll figure it out later."

Going down in the elevator, I asked what was going to happen next. "We'll let you know," Monkey said. "Keep your shirt on."

At the High Street subway station in Brooklyn, I looked back down the long escalator to see who was behind me. There was just a lady in a black overcoat with a red Macy's shopping bag. Above ground, I looked back once more just to make sure she wasn't following me. It was silly.

Before bed, I called Bob from my apartment. He wasn't in. I tried Michael.

"I'll call you back on my cellular," he said.

He rang me a minute later. "What about my phone?" I said.

"Your phone is fine."

"How do you know?"

"They're not tapping your phone."

"How can you be so sure?"

"It's extremely unlikely. Anyway, we're shutting down until we can find a new office."

"Is that what Monkey said?"

"Steak Knife. He and Monkey are trying to get to the bottom of this. Personally, I think Brandi's involved."

"Brandi?"

"Bob thinks so, too."

"That's ridiculous."

"She found the note."

"So what? I was with her. Besides, it would be a little obvious for her to find it if she was in on it."

"That's why it's perfect. Anyway, who knows? Maybe she's still pissed at Steak Knife for dumping her. Or she needs money to support her coke habit."

"She has a coke habit?"

"Yeah, you never noticed?"

"No."

"The way she's always going off to the bathroom. You haven't noticed?"

"She's got a weak bladder."

"You believe that?"

"Sure."

"Why are you so set on defending her?"

"I'm not so set. I just don't think she had anything to do with that note."

"Is it true, then?"

"What?"

"About you and her."

"Oh, for chrissake, Michael."

"Well, you shouldn't defend her so strenuously."

I started to say something more, but stopped myself. "What am I supposed to do now?"

"Wait. Somebody will call you."

Two days later it was Michael again. He had a new office phone number for me to pass on to my players.

"We're already set up?"

"Yeah, there are three phone lines in. The rest'll be up by Friday. You should call. They may need you."

"So that's it, we just move and now everything's okay."

"It's being taken care of."

"That sounds like wishful thinking."

"You can quit, you know. You don't have to keep working."

He was right. After I got off the phone, I hooked a beer out of the fridge, turned off the TV, which had been going silently during our conversation, and sat cross-legged on my new couch considering my options. The sensible thing was to quit. But it had not been a sensible job from the start.

In an odd way, the added threat of danger, which had a surreal quality, as it arose out of some words written on a piece of paper—and who knew better than I that they might be mere fiction?—worked to increase my interest. Of course I also had to admit to myself that this might be nothing more than an elaborate rationalization for the fact that I had started to make some money, that I had been touched by greed.

Whatever the case, I didn't quit. I called the number Michael had given me, and when I heard Monkey's frog-throated voice, I told him I was ready to come in when they needed me.

"Don't bother," he said. "We got too many people, too few phones. Take a couple of weeks off."

"But—"

"I gotta go. Bye."

I called Bob immediately.

"Don't listen to him," he said. "He doesn't know what he's talking about. It isn't going to be two weeks, so don't make any plans to go to the Caribbean."

"Well, what's the story?"

"The story is we got some people who are pretty shaken up by this thing."

Shaken up by a shakedown.

"Monkey is worried?"

"He won't admit it. But yeah. I'd say he is. This thing isn't over by a long shot. It's not just gonna go away."

I said that Michael didn't seem very concerned.

Bob said, "Let me tell you something. Michael has been very lucky so far. Nothing bad has happened to him in the entire time he's been doing this. I got a feeling the first time something does, he's going to be running home to Mommy. The fact is, I'm rethinking things myself. Don't tell anyone, but I interviewed at Goldman Sachs yesterday."

"You're kidding."

"It wasn't the right job, but if the right job comes along, I'm gonna take it."

"I think that's a good idea."

"You do?"

"Sure. I think that would be better for you."

"Well, you're not cut out for this, either. You shouldn't be doing this."

"I know."

"Why aren't you writing?"

"I don't know. I'm just not ready right now."

"Ready? What the fuck are you talking about, ready? You've already written shit. You've been published. You're beyond ready."

I laughed.

"Write a goddamn book, Curly, and make sure I'm in it if you want a best-seller."

I laughed harder.

"I'm not kidding. Are you gonna do it?"

"Sure."

"Good. Now, don't pay attention to Monkey. He doesn't know what the fuck he's talking about. I want you to come in the day after tomorrow. And bring your copy of *Anna Karenina*. There's a lot of shit in there I wanna discuss with you."

67

I began having strange dreams. In one I was with Anna, and three men with guns came up behind us. One of them put the cold metal barrel of his gun against the side of my neck and asked me if I knew who he was. I didn't, but I had the feeling I was supposed to and he would shoot me if I didn't come up with his name. Sweat began beading my forehead. Anna said, "Tell him. Tell him."

I woke up with my T-shirt soaked and sticking to my skin.

In another dream, also with Anna, we found a letter threatening to expose me. It said the whole world would find out what I was doing if I didn't quit. I would be the object of intense ridicule. I showed the letter to Pat and Monkey and the rest, and they said that Anna had written it. They found it funny that I would want to defend her. "Don't you get it?" they said. "It's so obvious. She has to be killed."

Real life had its own disturbing images. According to Michael, Monkey had arranged a meeting with the head of one of the tongs.

"What does that mean?" I asked.

"It means that we're on Chinese turf and they generally know what's going on in their own house."

"Monkey knows the head of one of the tongs?"

"Uh-huh."

"Jesus. So what are they going to do?"

"Help us get to the bottom of this."

"How?"

"If the note originated in Chinatown, they'll find out."

"And?" I could feel a grimace pinching my face.

"And they'll take care of it."

"What if this has nothing to do with the Chinese?"

"We're pretty sure that's where it stems from."

"I don't like the sound of this at all, Michael. I mean, what if it's a gang?"

"It's not a gang."

"Then what?"

"Just let Monkey and Steak Knife handle it. Okay?"

Oddly, about a half hour after our conversation, I got a call from Winnie. She was coming into town the next day. Would I like to have dinner with her and Michael, then go to a party afterward at her friend China's?

We met at Cucina di Pesce in the East Village. I purposely didn't bring up work in front of Winnie, since I didn't know what she knew. But pretty quickly she let it drop that she knew everything. It turned out that Michael had left the shakedown note lying around his apartment in plain sight—hello, Dr. Freud—and Winnie had seen it.

"So do you think this is all a big joke, too?" Winnie asked.

I looked at Michael, whose expression showed nothing.

"No, I don't think it's a joke, but I'm not really that worried," I said.

"You're not?"

"Mom, I told you," Michael said. "Monkey's taken care of it."

Winnie looked at me and I nodded.

"So it was this girl who was working in the office who was behind it?"

"She was part of it, apparently," I said, looking at Michael.

A while later when Winnie went off to the ladies' room, I said to Michael, "You told her it was Brandi?"

"We think Brandi was involved. Anyway, I fired her today."

I threw up my hands. "You fired her!"

"I know you think she wasn't involved. But the fact is, with things the way they are now we didn't have enough work for her anyway."

"That stinks," I said. "That really stinks. And you were the one to do the dirty work?"

"Monkey said I had to. It's my business, too."

"He should have done it himself."

"I wish he had."

"I mean, with Christmas coming up, and her and the kid going to go off to Australia .. . That's really low." I thought of her in the parking lot that day, asking me not to tell them we had found the note in her car, me assuring her that it would be okay.

"We were going to have to get rid of her anyway," Michael said. "She was making too many mistakes. She talks too much."

Winnie came back and said, "What's wrong?"

"Nothing," Michael and I said in unison.

After dinner we all went to China's party. Near two in the morning, after Michael had left and I had drunk too much vodka, I launched into an apparently cautionary rant to Winnie about what I thought was happening to her son and what the office was really like. Honestly, I don't know what I was thinking, and can surmise only that it had to do with my anger about Brandi. But I said Michael's heart was hardening, and what went on at the office was much worse than Winnie could imagine.

As any mother would have been, she was horrified. The instant I realized the effect of what I was saying—not as quickly, through my drunken haze, as I should have—I tried to retreat. It was too late. Winnie knew I'd been telling the truth. I finally changed the subject and went to get us more drinks. Maybe, I thought, sloshing Stoli into two plastic glasses, she was as fucked up as I was and by morning wouldn't be able to remember.

68

MY first day back at work I was uptight as hell, and it turned out I wasn't the only one. Everybody was jumpy. The new office was a crib that Monkey had been keeping in the East Fifties. From the looks of it, it had been his private fuck pad. At least it was furnished that way. Upper East Side Bachelor. Black lacquer cabinets and coffee tables, plush rugs, comfortable couches. Bob, Pat, and Spanky were manning the three phones—the rest still hadn't been activated—so I was assigned the task of charting for the night. Ed never charted, but it wasn't like solving Fermat's theorem. You just had to keep track of how much money was being bet on each side.

Before we got busy, I went off for ten minutes to take a shit; my stomach was acting up, which Bob said was psychological. When I came back, Pat said, "Have a good time? What were you doing, whacking off in there?"

"It sure is good to be back," I said. "I don't know what I'd do without you guys."

"You could always go see your girlfriend Brandi."

"Pat, that's not nice," I said. "Isn't it enough that she was fired?"

Pat shrugged. "Of course he's gonna stand up for her. He was fucking her."

"Really?" Bob said. "Is that true, Pete?"

"Yeah, sure."

"He was," Pat said. "All he had to do was put a little coke on his dick and she'd lick it right off."

Spanky laughed like a hyena at that.

Bob and I stayed late to work the extra half-shift for Monday Night Football. As soon as Pat and Spanky left, Bob said, "This is no kind of life. What the fuck are we doing here?"

I had no answer for him.

"I'm serious," he said. "I mean, look who we got here. We got one guy, Spanky, who's a fat, smelly slob with a bad attitude. We got another, Michael, who's deeply depressed and doesn't know it. We got Monkey, a gangster who kills people. We got Eddie, a drugged-out loser who needs to boss people around. Bernie, a fifty-year-old man who can't walk ten feet without stopping to catch his breath. And Pat. I don't even know what Pat's problem is because he's always so busy blaming it on everyone else."

"He does do that, doesn't he?"

"He's a fucked-up guy. But, hey, we're here with him. We're fucked up too."

"We are, aren't we?"

"Did you know that when I was twenty-two I made eighty-five thousand selling fax machines?"

"No way. You never told me that."

"Yeah. I was the company's salesman of the year. Then the crash came and wiped me out."

"You'd invested it all?"

"In stock options."

"But you still had the job."

"Yeah, I still had the job. The thing is, I hated it. I was good at it, but I hated it."

"You don't hate this?"

"This is different. I'm not proud of being a bookmaker. But I don't hate it. I don't feel like I'm dead. Know what I mean?"

"Yeah."

"I'm just not that proud to be alive."

69

IT was Anna who brought it to my attention. A story had run in Thursday's *Times* under the headline "Three Accused of Running Mafia Betting Ring." According to the story, one Andrew Garguilo —there was a photograph of him in the classic head-bowed posture as he was being led up the courthouse steps—a member of the Genovese crime family, had been the ringleader of a "super bookie" operation taking layoff bets from other bookmakers across the country. Charles Hynes, the Brooklyn D.A., was quoted as saying that bookmaking was "the cash cow" that ran organized crime.

I called Bob immediately after reading the story to ask him, as Anna had asked me, if that statement was true.

"It's bullshit," he'd said, essentially paraphrasing what I'd told her. "They don't control all of it."

"I just want to know what the deal is," I said. "Monkey meets with some Chinatown bigwigs ... I want you to tell me how he isn't connected."

"Look," Bob said, "Monkey has been around for a long time. He knows everybody. He hangs out with certain guys; then he goes and hangs out with other guys. The first guys think he's with the second guys, and the second guys think he's with the first guys. You know what I'm saying?"

"That I'm stupid?"

"I'm telling you, Pete, it's not what you think. The fact is, no one really knows who's with who, and they don't want to push too far to find out. It could be dangerous."

I didn't know if Bob knew and wasn't telling me or was simply in the dark, but later on at work—just me and Pat and Monkey—I couldn't resist asking.

"What about this guy Garguilo, who got busted?" I said.

"What about him?" Monkey asked.

"You know him?"

"Sure, I know him. I know him real good," Monkey said.

"Yeah, Andy Wilson," Pat said. "Lucky Andy."

"Lucky Andy? Wasn't that the name on the note?" I asked.

"Uh-huh."

"What does that mean? Was that note from him?"

Monkey shook his head. "That's what they wanted us to think."

"Why? Are we tied in with that guy?"

"Nah. We got nothing to do with him."

"But you know him?"

"What are you worrying about?" Monkey said. "Haven't you started making a little bit of money?"

Obviously not enough. When the bettors started calling, I was distracted and made mistakes. Before we left for the night, Monkey said to me, "I don't know what's wrong with you. You were so good at this when you started. What's happened to you? Your mind is on other things."

I listened without saying anything.

"You gotta pay attention to what's happening. A little bit of luck, you could make some real money doing this."

In fact, I pocketed another three grand that week. Which was chump change compared to what the office won: $350,000, according to Spanky. Despite the threat hanging over us, business was booming. And whatever misgivings Bob and I expressed to each other, we did nothing to get out. To the contrary, Bob cut a deal to take over the business of a small bookie named Ironhead, which further locked him in. The deal meant eight new sheets and about a hundred new customers to our office. After sewing it up, Bob swaggered around the office like a guy with a cannon behind his zipper. The deal forced Steak Knife to increase Bob and Michael's piece of the business to twenty percent.

Nothing more was said about Goldman Sachs.

70

THE fear and paranoia the note had brought on wasn't entirely snuffed out.

Nine shopping days before Christmas, I showed up at the new office and found Michael and Bob lying on the two couches, with the lights turned down low. No one else was there, and they barely stirred at my entrance.

"What's going on?"

"You want to tell him?" Bob asked Michael.

"No, you tell him."

"What?" I said. It sounded serious.

"We got another letter."

"You're kidding. What did it say?"

"It said that we had one more chance. Then they were coming in with guns."

"Are you shitting me? Where was it left? Who found the note?"

"Spanky found it. They taped it to his refrigerator."

"His refrigerator. Shit! They were inside his apartment?"

"Yeah. They said something about one of the clerks living in Brooklyn Heights, too."

"What!"

"The one with the curly hair and glasses."

"Are you serious?"

Michael couldn't hold it in any longer and cracked up. I looked at Bob. He was grinning.

"You assholes. What really happened? Something happened."

"Nothing happened," Michael said.

"Bob?"

"There was a letter. But it wasn't from them. It was from Michael's mom."

It turned out Michael had gotten a letter from Winnie saying he was either naive or a liar. She'd read the story in the *Times* and wanted to know how he could act as if we weren't mob-connected. "It was very sanctimonious," Michael said. "I faxed her back a really nasty letter in return. I don't think we'll be talking to each other for a while."

"What else did she say in her letter?"

"That she was trying to be a better person, more truthful. That she wasn't going to be a party to my secrets anymore. She was going to tell my grandparents what I was doing."

"So what did you write to her?"

"I said if she was going to start telling them the truth, then she should start by telling them how she did drugs in front of me and how she started taking me to bars when I was nine years old and how she let me stay out all night from the time I was fifteen. ..."

After work Bob had to meet his girlfriend. Michael and I had nothing going, so we decided to grab dinner. He seemed to want to talk more about his mom. "She told me I'm turning into a bully," he said as we got into a cab. I had a twinge of guilt. That was what I had said to her the night of China's party. Michael went on: "She said mob guys are bullies. So in my letter to her I told her she was a fine one to talk about bullying. She bullies everybody. Where does she think I learned it?"

Over dinner he said, "I've told her that we're not like some other guys. We've never hurt anyone or done anything like that. What we do is really no different from what guys on Wall Street do."

"Except that it's illegal."

"Right, but it's no more immoral than what they do."

"Which isn't saying much. It's all hard to justify. What they do, what we do—it doesn't contribute anything to the greater good. It's all about greed."

"Ninety percent of the population work at jobs that they hate," Michael said. "To me, that's worse than anything. At least I like what I do. I'll make a million dollars by the time I'm thirty. Then I can go do something else."

"Like what?"

"I'm not sure."

I wanted to say something to Michael, but I didn't know where to start. What could I say, after all? That I knew what it was like to be intelligent and stunted and to realize early on that the ordinary hell of a boring job wouldn't cure it? Maybe I didn't think I was as stunted as Michael. I could easily have been wrong, but it was clear I had no more clue than he did of how to go about making my life extraordinary and rewarding. I knew that when he talked about earning a million bucks it scared me. I knew you could probably hide the cold hard facts from yourself if you were making that kind of money. I was all too susceptible.

That night at home I counted up the cash I had squirreled away. There was nearly six thousand dollars hidden in half a dozen CD cases. I sat there looking at it for a while, a nice neat stack of crisp one-hundred-dollar bills. I'd paid off my credit cards, and it was good to be out from under. Even so, I felt oddly anxious after I put the money back, clicking the CD cases closed one by one. I tried reading Tolstoy but couldn't seem to concentrate. Finally I laid the book down on the couch, still open, and dialed Anna's number.

"It's me," I said.

"Hi. What's up?"

"Nothing ... I don't know ... I just felt like calling you."

"Are you all right?"

"Yeah. I'm fine."

"Are you sure?"

"Why? Do I sound like I'm not all right?"

"You don't sound great, sweetie."

"Mmm ..." I tried to explain about the money, about coming home and counting it and the emptiness I felt inside. "I guess I'm just really missing you," I said at last.

"Mmm," she hummed.

"Eight more days."

"I know."

Whatever natural wariness I'd felt about her plan to visit had faded, replaced by a growing sense of anticipation. "Is there anything you want to do? Any plays you want to see while you're here?"

"Not really. I mean, if there's something you want to see . .."

"I'll take a look in the paper."

"Mostly I just want to relax," she said. "I want things to be relaxed."

"Yeah, I know. God, I miss you, Anna. I think that's what's going on with me. I just really started missing you tonight."

She was quiet.

"I can't wait to see you," I said.

Instead of calming me, our brief conversation seemed only to intensify my longing. How could I make Anna understand the depth of my feelings?

On my way to work the following day, I went looking for a present for her. I decided I'd buy her jewelry for Christmas. A pair of earrings or something. But my attention kept getting drawn to engagement rings. It wasn't as if I decided I'd ask her and then went ring shopping; it was more as if the rings themselves were drawing me to the idea. The sparkling of the diamond solitaires in store windows mesmerized me, until I crossed the threshold of a jewelry store at Fifty-seventh and Third and a salesman saw me coming.

He was a thirtyish garment-district type who'd polished his act to the point where his Brooklyn or Queens accent was like the undercoat of a room that had been repainted in a different color. He had a coiffed mop of lustrous brown curls and sly brown eyes, and he talked quickly but with a wink in his voice—like, You and me, we understand each other.

"What are you looking for, something for your girl?" He adjusted his rust silk brocade tie and shot his cuffs.

I smiled.

"Hey." He shrugged. "I always give my wife jewelry. I mean, even if I wasn't in this business, I'd buy her jewelry. I just love watching her face when she opens the box. You know what I'm talking about."

I kept smiling and nodding.

"What were you thinking, specifically? Did you have a particular thing in mind?"

"I was just curious, really, about the price of rings."

He opened his hands, then clasped them together, almost as if he were praying, and leaned forward, bouncing the tip of his nose against his fingertips, smiling, looking up at me from under his eyelids.

"Now, of course, we're talking about something else. We're talking about something that goes beyond a gift. Something that's"—he nodded as if he understood now and there was no other word for it—*"big."* He opened his hands again and spread them, then bent over for a moment and came up with a midnight-blue velvet cushion, which he laid out on the countertop.

"Let me ask you," he said. "Does she have any idea this is coming?" He looked at me, his brown eyes moist with sincerity.

I felt embarrassed and trapped. A middle-aged blond woman glanced up from a wristwatch she was inspecting and smiled. To leave now would be an admission that I was not a serious person.... .

"The reason I ask," he said, removing several rings from the case and placing them on the cushion for me to inspect. "The reason is that if she already knows, I would ask you to bring her in, involve her in the decision."

"She doesn't know," I said.

"I popped the question to my wife over dinner at the Rainbow Room," he said. "She was so overwhelmed she had to get up and leave the table. It was great. It was the perfect thing. . . . Afterward you bring her in, have the ring fitted properly. But surprise is a nice thing. Very nice."

He swung the blue velvet cushion around so the diamonds were facing me. Most of the rings were too flashy and ornate, not at all to my taste or Anna's.

"I can see by your eyes these are not right," he said.

"I really want something very simple," I said.

"I know just what you mean." He bent over again and opened the display. "First off, give me an idea of your budget." The way he said it, I felt like I didn't want to come across as cheap. I said, "I don't know. Between one and two thousand?" He bounced his head from side to side with his lips pursed, as if weighing my words. "What about this one?" He put a ring on the cushion.

"Nice. How much is it?"

"This one is twenty-three hundred. It's a beautiful stone. Twenty-five points. The setting is twenty-four-karat gold."

I rubbed my fingers along my neck. "I'll have to think about it. That's a lot of money."

"We're talking about value, about something that's going to last. There are ways to get around cash-flow problems. We can work out some kind of installment plan. We can .. ."

He was off and running, no stopping him. Twice I tried to leave, without success. He kept enticing me back, first waiving the tax, then consulting with the owner and lowering the price a hundred, two hundred, finally three hundred dollars, to a flat two thousand. "You don't have to believe me," he said. "But it cost us eighteen hundred."

I almost did believe him. He had it working pretty good. I gave him a five-hundred-dollar deposit and told him I'd come by with the rest the next day.

Having made the decision, I didn't want to wait. At work I got Bob to lend me $1,500 until the next day—I wouldn't tell him what for— and I went back to the jewelry store afterward to get my ring. Exiting with my purchase, the silvery plastic bag in hand, I felt nearly high. It was such a strong and powerful step to have taken.

The feeling didn't last long. By midnight, alone in my apartment, I'd developed a case of buyer's remorse that turned into a full-blown anxiety attack. It was as if I'd gone temporarily insane. What had possessed me? I didn't know a damn thing about diamonds. I could have been buying paste for all I knew. The receipt said No Returns, and as I read that, I got a clutching knot in my stomach. What if I'd been taken? I'd need to get the ring appraised first in order to know.

But would I have any recourse then? Who knew? I kept taking the box out, opening it, looking at the ring, wondering one minute if I'd been taken and the next if Anna would like it, my anxiety taking shape around questions that were somehow not really to the point.

71

Turn on the TV!" It was Monkey on the phone, sounding very excited. "Tell Bob to turn on the TV! Channel seven."

I put my hand over the mouthpiece and told Bob to hit the remote. Spanky grabbed the clicker at the same time Bob did, and they wrestled over it. Monkey told me to put Bob on. I held out the receiver. With a disgusted grimace Bob gave up his tug-of-war with Spanky and settled for the phone.

The TV screen flashed to life. A Chinese man in a heavy leather coat was being escorted out of a police car. The camera lights glared harshly in the darkness of the street. The man in the coat put the back of his hand over his face and eyes.

"No shit?" Bob was saying. "That's him?" He held the phone away from his ear and said, "That's Dog Monkey."

"Yeah?" I said. Dog Monkey was one of Monkey's players. "Who is he really?"

Bob had the phone back to his ear and was saying, "He was arrested for murder? Yeah. Yeah. Wait a minute, let me watch this.. .."

On the TV the story had ended. The correspondent was signing off, saying "Reporting live from Manhattan South, this is Tim Fleischer for *Eyewitness News.*"

"You fucking talk too much," Bob yelled into the mouthpiece. "I missed the whole damn thing!"

He held the phone up over the small, round glass-topped table that was our makeshift workstation. Monkey's voice boomed out in miniature: "What do you mean you missed it? It was right there in front of your face! He was arrested for murder!"

We all cracked up. After Bob got off, I said, "Who is Dog Monkey?"

"Henry Lum."

"Who's Henry Lum?"

"The guy Monkey talked to about our problem."

"Wait a minute, he's the guy? He's the head of the tongs?" Monkey had told us only a couple of days before that our problem had been taken care of.

Bob said, "Sure. You didn't know that?"

"And he was just arrested for murder?"

"It's for something else. The murder doesn't have anything to do with us."

"How do you know? Did Monkey say that?"

"He said it was something else."

"So what you're telling me is that this guy—this Henry Lum who took care of our problem and has now been arrested for murder—probably just scolded whoever wrote us that note?"

"How should I know what he did? He took care of it."

I let the matter drop. Whatever the point of my asking, at heart I did not want an answer.

The next day—and whether it had anything to do with Dog Monkey's arrest or not, I don't know, for there was no further discussion of the subject—we moved back into the old office on St. Marks Place; we'd been paying Krause a monthly retainer while we were gone. Bernie stayed at the Upper East Side office to relay calls.

Aside from a cheap brown carpet remnant that had been laid down to appease the guy on the floor below, the only difference in the old dump was the absence of the cave dweller himself, Morry Krause.

"Where is he?" I asked.

"Cape Cod," Michael said.

"What's he doing there?"

"Writing a book, supposedly."

"Yeah, sure," Pat said. "That'll be the day. I'll tell you what he's doing. He's sitting on his fat ass with a clicker in his hand watching *Black Chicks with Dicks*. Go check his room out. You ever look at his smut collection? He's one sick fuck."

"Pete went to the Cape to write a book. Didn't you?" Michael said. Spanky snickered.

"Thanks, very nice of you to bring that up, Michael."

"How much time did *you* waste up there?" Pat sneered.

"What about showing a little Christmas spirit around here?" Bob said. "Let's all be nice to each other for once."

"Fuck Christmas! I hate fucking Christmas!" Pat said.

"Shut up, everyone," Michael said, putting his hand over the mouthpiece of his phone.

The others looked at him, respecting the tone in his voice.

"Great," he said, hanging up the phone. "That was Steak Knife. The Nickel office just got popped."

The Nickel office was one of our most frequently called outs, and everyone got into an immediate sweat about the bust, saying, "Let's get the fuck out of here." My heart was pounding.

But Bob, in his gruff baritone, said, "They couldn't possibly know we're back here in Krause's place. Besides, if they were going to bust us, we'd be busted right now. They coordinate these things."

Pat decided to take a stroll outside the building anyway to see if everything looked okay. Before anyone could object or agree he was out the door.

"What a sphincter hole," Spanky said.

Bob said, "Let's tell everyone we're closing at seven-thirty."

We continued fielding calls, but it was edgy, everyone very uptight. Then I got a call from a guy saying he was Rascal. Only we didn't have a Rascal and his voice didn't sound familiar to me. I asked him if he was sure that was his right name, and he said, "Yeah." Spanky took the phone from me and asked him the same question, then told him he must have the wrong number and hung up.

After that we said to hell with it, let's get out. We were spooked.

The hall was quiet. We tiptoed down the stairs, peering around corners. Outside the building, there was no sign of Pat. But everyone we did see looked like an undercover cop, even the homeless guy sitting in front of the neighboring building. I shivered in the thirty-degree air. In our rush to get out, we'd left our coats behind.

"Four guys coming out of this building in shirtsleeves. We don't look too suspicious or anything," Spanky whispered to me.

Without saying anything more, we split up and walked off in different directions. I got to the corner of St. Marks and Avenue A, where there was a pizza parlor, and waited for Bob to catch up.

"Want to get a slice?" he said. "As long as we're here."

Once we got past the initial paranoia, the overwhelming feeling was one of foolishness. The street suddenly looked as it always did, crowded and innocuous. Outside, among the regular citizens—a relative term in Alphabet City—we felt as if nothing could happen to us. Looking down the block, I saw that Spanky and Michael had arrived at a similar conclusion and were drifting back toward Krause's building. Bob shrugged at me and we followed suit. It was seven-thirty by then; if the cops were coming they would have come already. We had to go back to get our coats anyway.

The office was eerily quiet, the phones not ringing at all. "Why don't we just call it a night?" Michael said.

Spanky and I looked at Bob hopefully. But just then one of the phones rang. I picked up. It was one of our biggest customers, Dodge. "Where you guys been?" he said. "Busy cleaning the brown stains out of your underwear?"

I wrote up a ticket he wanted on an NBA game. Brown stains in our underwear. Man. News traveled fast in the gambling world. Scarily fast.

72

WE left the office at ten to eight. Bob and I went to the 57 to watch
Michael play in a pool tournament. I called Anna around eleven
o'clock, and her machine picked up. If she'd gone out, a babysitter
would have answered. When I got home atone in the morning, I tried
again. Still no luck. I tried to convince myself she'd gone to bed early
and turned off her ringer.

I left another message at eight in the morning.

Just after eleven a.m., she finally called back, talking in a rush. "So
where were you?" I asked, trying to keep my voice as light as possible.

"I went to Richard and Lily's for dinner. Then I wound up staying
there. Nathaniel had a sleepover at a friend's and I was kind of drunk,
so it made sense."

"Did you have fun?"

"Yeah, I did."

"You sound a little funny. Are you all right?"

"I'm just really tired, I think."

"I'm getting the feeling here that maybe I should be jealous. Should
I be jealous?"

"What do you mean?"

"You know what I mean."

I heard her take a deep breath. Damn.

"I'm not thinking clearly enough to have this conversation," she said.

"So that means I should be jealous."

"You know, young girls they do get woolly. I mean weary."

"You mean you did sleep with someone? You really did?"

"Oh, Pete."

"No, it's fine. I just can't believe the timing. I mean since you're seeing me in a week."

I could hear her breathing, groping for the right response. "Ugh, I hate this," she said finally. "I hate hurting you."

"*You* hate it?"

Silence.

"You don't even know how much I was looking forward to seeing you," I said.

"Do you not want me to come now?"

"I don't know. . . ."

For what felt like minutes, we sat there, a thousand miles apart, each of us waiting for the other to speak.

"This is .. . This is really . . . Did you just meet this guy?"

"Does it matter?"

"I mean I just don't understand, Anna. ... I thought things with us were . . . We were talking so well." I couldn't get over her timing. Didn't she see anything funny about it? Didn't she consider that maybe she had committed an act of sabotage?

She kept saying these things just happened. But I said I didn't think so. Not with her.

Still, I had trouble getting mad. It would have been clear to anyone by then that she was incapable of giving me what I wanted. So why each time did I expect different results? Out of what character trait was such optimism born? Was it stubbornness or stupidity? I realized, with some horror, that this was not merely a pattern of my love life; I had done the same thing in my writing. Was I persistent in the way I went about things or just dense? Hopeful or denying?

I remembered a friend saying to me once, "Your problem is you don't know when to let go of a bad idea," and how the truth of that still stung.

I drew a deep breath. "I think I can't talk to you anymore," I said to Anna.

Her voice got small and quiet. "What does that mean?"

"It means that I'd like it if you don't call me."

"Are you going to call me?"

"I don't know."

She chewed on that.

"But what if I move or something happens?"

"If you move, let me know. Otherwise please respect what I'm asking."

For several minutes, after severing the connection, I sat on the couch immobilized. Then I had myself a good cry. I could have forgone the tears, I suppose, but I helped them along, pushed them, thinking about her. I wanted the cathartic release, the literal sense of watershed. Afterward I felt oddly okay, though I knew the feeling was illusory, temporary. Real sadness still lurked around a comer somewhere, and I had no idea how it would feel when it found me. Or whether I was willing to let it.

73

As if the holidays weren't rough enough, I wound up having to work on Christmas Day taking calls from all the degenerates out there who had even less of a life than I did. Just three of us came in for the two NBA games: me and Bob—the two Jews—and Spanky, the heathen. Also Krause, who had returned from the Cape without having written a word. He seemed in pretty good spirits despite that, and when he came out of his cave in a T-shirt and soiled Skivvies to say hello, I was almost happy to see him in a Salingeresque "I miss 'em all, every goddam one of 'em" kind of way.

Krause started going around the table, calling us by the nicknames we brought to mind. I was the Professor. Bob was the Oaf. "And then," Krause said, when he got to Spanky, "there's Sasquatch."

Spanky made a sour face. But Krause wasn't finished. He said, "Better call Leonard Nimoy, *In Search of. . .*"

Bob and I grinned.

Spanky was haughty, as if Krause's insults were beneath him. "Why don't you shut up, you fat piece of shit?"

"Ooh, the creature talks."

"You looked in a mirror lately, Morry?"

"Yeah, but at least I don't see what you see"—Krause gave his best actor's pause before delivering the punch line—"the two ugliest people in the East Village."

Again we laughed and Spanky tried another lame comeback. "You're just jealous because I can still get laid, old man."

Krause dismissed him with a snort. "You couldn't get laid if you walked into a women's prison with a handful of pardons." Bowing with exaggerated elegance, he returned to his cave triumphant.

Afterward Spanky fell into a foul, venom-spewing mood, most of the poison directed at me. That was okay. I was actually happy to disgorge some of my own venom. But Bob got sick of the insults. He said, "What the fuck is with you, Pete? I understand this shit coming from Spanky, but this isn't like you. Isn't your girl in town? You should be in a good mood."

"Don't even talk to me about her."

"Uh-oh. What happened? Don't tell me she blew you off?"

I groaned and waved off his question.

"Right before Christmas? Oh, man. I'll bet you bought her some expensive gifts, too, didn't you?"

"You don't even know. I was going to ask her to marry me."

"You got her a ring?"

Spanky, predictably, found this part hilarious.

"Aren't you glad to be spilling your guts out in front of us?" Bob said.

"I don't care. I'm so fucked up right now I don't care who knows. The worst part is that I'm stuck with the damn ring. I can't return it."

Spanky practically spit, he was laughing so hard.

I shot him the finger.

"How much was it?" Bob asked.

"Don't ask." I looked at a patch of the wall where the paint was flaking off in particularly large sheets. "Two dimes."

"Two dimes! Holy shit! You have that kind of money to throw around?"

"I wish."

"Well, bring that ring in tomorrow. I'll take a look at it. Maybe I'll buy it from you."

"What, to give to Stacy?"

"I'm thinking about it."

"How come your girl broke up with you anyway, douche bag?" Spanky asked me, his tone verging on tender.

"'Cause she's a cunt," Bob said. "Why do you think?"

"Maybe she has a problem with what Pete does for a living."

"If she knew what kind of morons he did it with, she'd definitely have a problem."

"Fuck you. We can't all be Ivy League graduates," Spanky said.

"Enough, you guys. Enough, please," I said.

"Pete's right," Bob said. "Let's all be nice to each other. Let's see if we can make it through the rest of the session without being nasty."

That night I went to Christmas dinner at my aunt and uncle's place in Brooklyn, thankful that I had a family, a place to go to, even if my cousins' talkative high spirits and news of recent achievements made me feel like the poor relation.

Only afterward, returning home to my empty apartment, did real sadness finally begin to catch up to me. I sat on my couch, staring at nothing, feeling the oppressive crush of the too-near walls and breathing in the hot, hissy air of the old steam radiator. After a while I began thumbing through my address book, looking for someone to call. I came to the M's and skipped over them, not wanting to see Anna's name. It was so difficult for me to hold on to the anger.

I breathed deeply, putting the address book down. The tears started slowly, then came hard. I gave in to them finally, the delicious release of them. Getting up, I went into the bathroom, pulled the string on the overhead bulb, and looked at my contorted face in the paint-speckled mirror. It was like watching someone else. I stared at my reflection with odd fascination, crying even harder. Then I stopped suddenly and grinned. Yellow teeth, glittery blue eyes, big crooked nose. Who the fuck was that face in the mirror?

I had begun sobbing again when the phone rang. I wiped my eyes, heart pounding, and went to pick it up.

"Curly." It was Bob.

"Oh. Hi."

"You thought it was her, right?"

"No, I don't know what I thought."

"Well, forget about her, do yourself a favor. She's a cunt."

"Is that what you called to tell me?"

"No, I called to tell you not to bother bringing in that ring tomor-row."

"You changed your mind?"

"Something like that.... Stacy dumped me tonight."

74

I'M sick on top of everything," Bob said. He threw his gym bag on the couch and slumped into the chair by the first phone.

"If we're not careful," I said, "we're going to end up walking around like Paddy, saying 'I hate fucking Christmas!' "

"The only good thing that happened yesterday was that the office made fifty grand. Which reminds me . . ." Bob threw me a sealed envelope.

"What's this?"

"A little extra."

I tore the envelope open. There were two C-notes inside. We'd gotten our real Christmas bonus a week before, but Bob explained that this was a small additional token. I wasn't about to complain.

"Just don't tell Spanky about it," he said.

"I won't mention it. Is it just you and me tonight?"

"Yeah, this is a nothing shift. I could've worked it by myself. There's the one football game tonight, and it's for degenerates only."

I was glad to have the shift, not for the money but for the company. I had no idea what I'd have been doing otherwise. Watching TV at home, probably.

"Curly?"

"What?"

"I'm fucking sick as a dog." Bob honked into a wad of toilet paper he'd torn off from a roll beside him on the desk.

"It's psychosomatic, Bob. You're bummed about Stacy."

"Nah. It's not about her."

I didn't believe him, but I wasn't going to argue. He'd been vague when I asked him why she'd broken things off. "A buncha reasons" was the most I could get out of him. I knew one of the raw nerves was that Stacy's father had a gambling problem, so in Bob there were a lot of uncomfortable echoes for her.

To make matters worse, Bob had secretly been letting Stacy's old man place bets with the office under the name Dad for Topsider. We had all told Bob that he was crazy to let the old man play, but Bob had shrugged it off. "He wants to bet," he said. "I'm supposed to tell him no?"

Monkey had agreed: "A man wants to bet, you let him bet." Maybe Stacy had found out. Or maybe it just came down to the fact that she didn't want to marry an outlaw gambler.

Bob blew his nose again. "We gotta do something, Curly."

"What do you mean?"

"I don't know. Just do something. With our lives. Don't you get scared?"

"*Get* scared?"

"I mean, you could wind up like him." Bob gestured toward the bathroom, where, behind the closed door, the shower had been going since we arrived. "Hey, Krause!" he yelled. "You drown in there?" No response. Bob wagged his head. "I think he's drowned. He's been in there for a fuckin' hour."

"Maybe he's trying to make up for all those showers he missed," I said.

The sound of the water finally stopped. I knew Bob wouldn't be able to resist taunting him. He yelled Krause's name.

"What the fuck do you want?" Krause said when he came out of the bathroom a few minutes later, sticking his wet sheepdog head around the corner.

"How was your Christmas?"

"You got me out of there to ask me how my Christmas was? Why don't you ask your cheap fucking boss?"

"Hey, I didn't see you giving us anything."

"I'm not supposed to give anything," Krause said. "I'm supposed to get." He hung in the doorway another moment, then withdrew. About eight o'clock, as Bob and I were sorting the night's work, he reappeared.

"Whoa," Bob said, looking him up and down over my shoulder. "Check it out."

I turned. Krause had put on a clean blue shirt and khakis. His thinning light brown hair was combed back carefully in a grease-era pompadour, and he had splashed on liberal amounts of Old Spice.

I whistled.

Krause moved closer to Bob and, from behind his back, produced a boxed but unwrapped videotape. "Here."

"For me?" Bob asked, taking the offering.

Krause gave a modest twitch.

"Look at this!" Bob said, holding up a graphically illustrated video-cassette entitled *Butt Sluts*. "Wow! Is this part one or two?"

"You making fun of me? You making fun of me, give it back."

"No, no. I'm serious. 'Cause I already have part one."

"Fuckin' asshole. Here, I got one for you, too," Krause said, handing me a cassette called *Anal Agony*.

"Mmmm," I hummed appreciatively. "Thanks."

I looked at Bob and tried to keep a straight face. Krause seemed antsy, glancing at his watch nervously. "It's past eight," he said.

"So?" Bob said.

"So I got someone coming over and I don't want her to see you crim-inals—all right?"

"What are you talking about, Krause? We leave when we're finished. You know how it works."

"This is my fucking apartment!"

"I'm sorry, Krause. But that isn't our problem."

Krause hovered while we taped up the envelopes and collected our stuff, including our gifts. We were at the door when the bell sounded. Bob and I went down the hall and waited by the elevator. Krause rang the buzzer, then came and stood in his doorway, staring daggers at us.

"Can't you lazy fucks even walk down the stairs?"

"No," Bob said.

The elevator door opened a moment later, and this grotesque-looking black chick in a blond wig and brass-colored lipstick stepped out. Bob and I could barely contain our laughter. As soon as the elevator door slid shut behind us, we exploded.

"No wonder he wanted us out of there."

"Oh, man."

"God, can you imagine fucking her?"

"Can you imagine fucking him?"

I had to lean on Bob to keep from sinking to the floor of the elevator. But by the time we reached the ground floor something else had hit me: "What are *we* laughing about?"

We looked at each other, tears of mirth still glistening in our eyes.

"I mean, what are you gonna do when you get home tonight?"

"I don't know." He shrugged, sobering up. "What are you gonna do?"

"The same thing you're gonna do."

"Pop in that tape?"

"Pop in that tape." I nodded. "Exactly."

75

NEW Year's Day the whole crew worked the bowl games: Bob, Michael, Spanky, Eddie, Bernie, Pat and a new guy, Sal, a cousin of Steak Knife's—everyone but Monkey, who was with his wife's relatives in New Jersey.

It was the first time we'd had a full crew together since before Christmas, and almost everybody was glad the holidays were over. "To hell with families," Bob said. "This is our family."

I'd been trying to hold on to my anger toward Anna. I kept thinking if I held the forces of sadness at bay long enough, maybe they'd simply go away.

I also realized, sitting at the big oval table in Krause's apartment on the first day of the new year, that I had grown fond of this sorry group of guys, that I had grown fond even of smartass Spanky in his smelly red Polartec pullover and backwards ball cap, and puffed-up, insecure Eddie in his handmade shirts and lizard-skin boots. I felt as if they had accepted me in their way, didn't care what I was, or what I had or hadn't done.

At halftime of the Sugar Bowl, the phone calls tapered off for a while, and when a deck of cards materialized, Bob, Spanky, Bernie, and I started a round of Indian-head poker, the game where you hold a card up on your forehead and bet that your card is higher than the cards you see on the other players' foreheads. It's a ludicrous game, the opposite of regular poker: you know everyone else's hand but not your own. But bluffing is still the key, and we were giggling like crazy as we played, which was driving Eddie—who wouldn't lower himself to play —crazy.

Then on one deal Spanky put up a two on his head, Bob put up a two on his head, Bernie put up a three, and we all started betting like mad, which made me realize I must have a low card, too, but of course it couldn't be as bad as theirs. Pretty soon we had bet the pot up to several hundred dollars, and I was thinking, Christ, maybe I have a two also! We finally declared a halt and put down our cards, nearly crying—the more so when we saw our cards: two twos and two threes. Like Bernie, I had a three, which meant he and I got to split the pot. Bernie used up so much oxygen laughing that he had to reach for his inhaler, and Spanky was so pissed he flung a pencil at him. When Bernie retaliated, he missed Spanky with the No. 2 and hit Pat in the head.

The whole table broke up. Pat jumped to his feet and said, "You think that's funny?" staring across the table at Bernie and clenching and unclenching his fists.

Bernie said, "Siddown, Pat. It's nothin'."

Pat rubbed his head and said, "Nothin' for nothin'." He looked at his hand and there was blood on it, the sight of which crippled us.

"Very fuckin' funny, you motherfuckers," Pat screamed. "I'm bleeding here like a stuck pig."

We all had to look down at our sheets and concentrate very hard. But Pat had seen enough. "Fuck you all. I'm out of here." He stormed toward the door.

"Pat, c'mon," Bob said.

"Pat, I'm sorry," Bernie said. He cupped a hand by his mouth and under his breath added, "Sorry I didn't hit him in the eye."

"What's that?"

"Nothing, I said I'm sorry."

"You won't be laughing when I come back with a gun."

"Pat, c'mon. He said he's sorry."

Pat didn't feel like listening. He slammed out the door, leaving us momentarily shocked and silent. It didn't last long. Soon we were mimicking him mercilessly. " 'You won't be laughing when I come back with a gun,' " Spanky imitated. " 'Nothin' for nothin', Bernie. I'm bleeding here like a stuck pig.' "

No more than ten minutes after Pat's dramatic exit, there was a loud crashing thud at the door. We all turned, half smiling, ready for his cranky return. But there was another booming thud that dried the laughter in our throats.

"That ain't Pat," Bernie said.

"What are you—"

"That's them."

There was one more crashing thud, and I sprang to my feet, heart pumping. I moved right, then left, and went nowhere. The door burst open and half a dozen cops in bulletproof vests, guns drawn, poured into the room.

"I can't fucking believe it," Eddie whispered.

I sat back down.

"All right, everyone, hands on top of your heads," said the lead cop, who had receding ginger-colored hair and a bushy mustache. "Happy New Year, fellas."

"I can't believe it," Eddie muttered again.

A short, wiry black cop started going around the table, taking our phones off the hooks, some of them in mid-ring. I could hear tiny voices coming out of them. "Hello? . . . Hello?"

My heart was throbbing, my mouth tasted like paste. I looked straight ahead.

"You guys know the drill here," the mustached cop said. "So let's make it easy for all of us." Another cop in a bulletproof vest came in with a Polaroid camera. The lead phone, over by Michael, had begun its nerve-jarring off-the-hook beeping. The short black cop told Michael to unplug the handset. Michael did as he was told. My phone began beeping next, and I took one hand off my head, very slowly, and began reaching forward.

The black cop spun and jumped, his gun, held in two hands, pointed right at my chest. "I'll spray you all over that wall! I swear it!" he yelled at me. "You move again, I'll spray you all over that wall!"

I put my hands back on top of my head. My heart felt like it was pounding in every part of my body.

"Don't let Ricky bother you," said Mustache, much amused. "He just came off a six-month rehab for shooting someone. He's a little jumpy."

A cop wearing a Jets cap came out of the bedroom and said, "Nothin' in there."

Bob and I exchanged glances, and I knew just what he was thinking: Bad enough that Pat had gotten away! But Krause? What were the odds of him not being there?

After we were cuffed, we were led out of the apartment into the hall. A girl from the apartment at the end cautiously poked her head out. "What's going on here?" she asked. She was tall and blond. The cop in the Jets cap said, "Go back in your apartment, ma'am." She turned back inside and I heard her say, "It's the guys from next door." Then there was another, shorter, girl in the open doorway, looking at us.

"What did they do?" the shorter one asked.

"Please go back inside your apartment," the cop said. "This is police business."

"Is it drugs?"

"They're bookmakers."

"Really?" the tall blonde said.

"They seemed like such nice guys," the short one said.

"It's just a mistake," Bob said to them. "It's all a big mistake."

The tall blonde looked at us, then back at her roommate. They giggled.

"Go back inside," the cop said.

They shut the door.

"I always knew she liked me," Bob said.

"You gonna make your move before or after prison?" I asked.

"Shut up," the black cop, Ricky, said.

Mustache sauntered out of the apartment, looking at us with a gleam in his eye. "Well, well, I see a couple of familiar faces here. Bernie, aren't you getting a little old for this? Gonna make your poor wife worry when you don't come home." He shook his head.

"Ahh, she won't worry," Bernie said. "She'll be happy to have me out of the house for a night."

"Who's that?" Mustache walked down the line toward Eddie, who was wearing jeans, Tony Lamas, and a shirt with the ten, jack, queen, and king of hearts and the two of clubs embroidered on the pocket and the ace of hearts on one cuff. "Eddie Brudney! Alex's little brother."

Eddie shrugged sheepishly.

"We'll have to catch up later. In the meantime, let's play a little game. Let's see what we've got in our pockets." Mustache moved back up the line. I could hear the rest of the cops inside the apartment, turning stuff over, throwing it into boxes. "We'll start over here," Mustache said, stopping in front of Bob. He reached into Bob's back pocket and lifted his wallet. "Nice wallet," he said. "Perry Ellis? Mmm, look at this." He extracted a thick wad of bills. "Let's see what we got here." He licked his fingers and started counting with the professional dexterity of a bank teller or cashier. "One hundred, two, three, four, five, six, six-twenty, forty, sixty, eighty. Six hundred and eighty dollars. Not bad. Not bad."

He put the money in a brown envelope and started going through the rest of Bob's wallet. He pulled out Bob's Dartmouth I.D. "Hey, Ricky, look at this. We got ourselves an Ivy Leaguer. Too many fucking lawyers and doctors, right? Let me see, think I'll go into bookmaking." We were all snickering now. The more Mustache talked in his mile-a-minute New York voice, the more he reminded me of some actor or comedian I'd heard. Rob Reiner? Billy Crystal?

"You guys taking action on the Ivys? I mean, since you got an expert on premises, why not?"

Bob fidgeted.

"Me, personally, I wouldn't play the Ivys with a gun pointed at my head. I make it a practice not to bet on schools that spend more on their libraries than they do on their football stadiums. And who knows with these kids? Maybe they got a Chaucer paper due or a big exam coming up. Or they sprained their IQ. Who the fuck knows?" He continued to poke through Bob's wallet. "Let's see what else we got here. MasterCard, Visa, American Express. What is this? You guys don't use cash anymore? Don't you know you're bookmakers? You don't need credit cards."

"You gotta have a credit card to rent a car," Bob said.

"*Rent?* What, rent? You mean to tell me you don't have a Beemer parked outside? Maybe you mean for when you're vacationing in Aruba?"

Bob smiled. He and Michael had gone to Aruba the previous winter.

Mustache moved up the line to me next. Took out my wallet, counted my money. I had only a hundred and seventy in cash. "I'm sorry to tell you, Pete," Mustache said, reading my name off my driver's license. "But you're running a distant second here in the money department." He reached into my shirt pocket and removed a pink betting slip. "Curly," he said, reading it off. "Indy minus six and Kansas minus eight in a ten-time parlay." Everyone in the line was laughing.

"Curly, huh?" Mustache said. "Well, let's see, Curly, I kind of like Indy, but Kansas, I dunno. You seen them play? They're up-tempo. They're not going to like playing against a methodical team on the road. And in a parlay? Shame on you. Don't you know that's a sucker bet?" He had my co-workers—my co-prisoners—howling now, and he stuck the ticket back in my pocket, saying, "Curly, I think I'll let you keep this. Maybe they'll cut you a break and say it doesn't count because of the bust."

Next he did Eddie, who had $1,000 in cash on him. "I think," he said as he started counting the big roll, "we have a new leader here. Yes, we definitely have a new leader! Congratulations, Ed! One thousand dollars. A dime. Two nickels. Two hundred times." He put the money in another brown envelope. "You won't miss this, will you, Ed? This is tip money for you."

Mustache walked back up the line, stopping at Spanky. "And who do we have here? Let's see." He lifted Spanky's wallet, an embroidered Guatemalan thing, and opened it up. Two fives and a one. He shook his head sadly. He looked at Spanky's driver's license. "Irwin Hall."

"Irwin!" Bob snorted.

"Irwin?" the rest of us repeated.

"Well, the night is full of surprises, isn't it?" Mustache said. "See, this kind of thing happens and you find out all sorts of stuff about each other. Let's see what other surprises Irwin's got for us." He reached into Spanky's pocket and pulled out a huge wad of cash. "Uh-oh. Eddie, you're in trouble. I think we just hit the mother lode here. Irwin's got a lot of cabbage." Mustache started counting, went through seven one-hundred- dollar bills, then came to the twenties. He sighed and kept counting. "Seven-twenty, forty, sixty, eighty; eight hundred. Eight-twenty, forty . . . Irwin, where I am going with this? Help me out here."

Spanky sighed. "Sixteen-forty."

Mustache kept counting anyway. "Sixteen. Sixteen-twenty. Sixteen-forty it is. We *have* a winner!"

He sealed the money away and said, "All right, guys. I gotta go see what's happening inside, so I'm going to leave you in the care of Ricky here. Ricky, try not to shoot anybody, okay?"

We waited out in the hall while the crew of eight cops cleared out our office, removing boxes and black Hefty bags full of phones, records, and equipment. On one trip, Mustache came out of the apartment carrying one of our melamine marker boards. It had all the college basketball games and lines neatly penned in Spanky's—Irwin's—handwriting. "Marker boards," Mustache said. "Very professional." He looked over the lines, smoothing his mustache thoughtfully. "I'll tell ya, you know who I like tonight? I like Auburn. I know it's a lot of wood to lay, but Tennessee's a tired team."

"Go ahead," Bernie said. "Your money's good with us."

It took them half an hour to finish clearing out the office. Finally they took us downstairs and out into the rain, put us into four separate unmarked cars, and drove us across the bridge to the Brooklyn District Attorney's building. We were taken to the sixth floor, led down a few hallways cluttered with Xerox machines and boxes, and told to sit on two benches outside the detectives' room.

We were there for hours while the paperwork was completed. A fair amount of our conversation was devoted to Pat and his miraculous escape. If it hadn't so obviously been a planned operation, we would have been convinced that Pat in his rage had tipped the cops. Bernie said, "Let's give 'em Pat's address. It ain't right that he ain't here with us."

Meanwhile, other bookies busted in the same roundup were being brought in and paraded past us into other waiting areas. The contrast between us and them was striking. They *looked* like bookies. They wore bright-colored silk and nylon Sergio Tacchini and Fila warm-up suits unzipped to reveal chest hair and gold chains. They had intense black eyes, pockmarks, hair plugs; one guy had a cast on his right hand. Real *cafones* or *gindaloons,* as Monkey would have called them.

Bob and I exchanged looks. Finding ourselves in the same fix as these thugs had us thinking the same thing. "You know," I said, "if we're such smart guys, how come we're here?"

"I know," he said.

"It's ridiculous. Monkey and Eddie and Bernie say this comes with the territory, that if the cops want to pop us they can do it anytime. But I don't believe that. If you're going to be in this business, you have to find ways to avoid this."

"That's what I've been telling you guys all along," Spanky said. "Isn't that what I've been saying?"

"Yes, Irwin, that's what you've been saying."

Bob and I cracked up. *Irwin.*

"No, really," Spanky said. "If Steak Knife and Monkey weren't so fucking cheap, we could do call forwarding and use cellulars."

Bob agreed. "There's definitely a way to beat this stuff. I'm not sure that's it. But there's definitely a way. We just need to find a guy who's an expert on phone security, who can tell us exactly what techniques the cops use and how to outsmart them. It might cost money, but it'd be worth the investment. I mean, this will wind up costing us at least a couple of hundred thousand in lost business and expenses."

"Exactly," Spanky said.

We all nodded in agreement. Next time would be different.

Next time?

Fingerprinting took several hours. They had one schmucky detective doing all twenty-three bookies pinched in the sweep, and it took him about fifteen minutes per person. It was a strange process. When it came my turn, he inked a plastic palette with a roller, took my thumb in hand, and after telling me to relax, rolled it from side to side in the sticky ink. Then he took the wet black thumb and repeated the rolling motion on a white card with spots marked for left thumb, middle finger, right thumb, and so on. There were three white cards to fill in, and the process was repeated with every finger of both hands three separate times. It made me feel like a child to have my hand taken by someone else and manipulated. Afterward I was given a Wetnap to clean up with. And I was told I could make my phone call. Bernie had already talked to Monkey, who was going to call Steak Knife, who would arrange for a lawyer for us. I couldn't think of anyone to call. Certainly not my parents or Anna. I ended up calling my machine. "Hi, it's me," I said, under the watchful eye of the detective. "It doesn't look like I'll be coming home tonight. So don't wait up."

The detective replaced my handcuffs with little regard for my comfort. He couldn't have cared less who I had at home.

At 2:30 a.m. the waiting on the benches ended. I was cuffed to Eddie, Bernie to Bob, Sal to Michael, and because of our odd number, Spanky was cuffed behind his back again. I was glad that something was happening finally. I was sick of that hallway. But Bernie said, "You're going to be wishing you were back here pretty soon. Believe me."

76

THE Brooklyn House of Detention was a short ride away. We were escorted out of the cars and through the hard rain down a steep, bending driveway to a side entrance. By the time a cop came to the caged window of the door, eyed us, and decided to let us in, we were soaked. The place was gloomy and narrow inside, like the underbelly of a football stadium. We were led through a couple of dimly lit corridors to an area where our pictures were taken, then through a heavy locked door, down another corridor, and into the first level of the detention area. Immediately we were herded into an eight-by-twelve-foot cage where several of the other bookies arrested with us were already waiting, along with some other nasty-looking characters. One of the other bookies, a chisel-faced thug in Sergio Tacchini, kept looking at me with his wide-open unblinking yellow-brown eyes. He gave me the willies. Spanky was next to me, and I whispered in his ear: "The guy in the Sergio Tacchini has definitely killed people."

Spanky looked at me like I was an idiot and told me to shut up.

"I'm telling you," I said.

"Just shut up, Pete." He was pissed because the cop who'd ushered us into the cell had refused to cuff his hands in front of him. His shoulders ached, which is what happens when you are cuffed in back for any length of time.

There was some shuffling of prisoners in and out of our cell. Bob and Sal and Bernie and Michael got moved somewhere else, as did some of the other bookies. For a while I was glad that I was cuffed to Eddie and not alone like Spanky. It made me feel better knowing I wouldn't be left by myself. But then Eddie struck up a conversation with some Rasta man in a Ja Love T-shirt and a torn parka, and before my disbelieving eyes Eddie started making a drug deal—Mustache evidently hadn't taken all of his pocket money. We were ten feet away from about fifty cops, and Eddie was buying smack. "I don't know you," I said, trying to separate myself from him as much as possible despite the fact that we were linked by unbreakable steel cuffs. Eddie scoffed at me, palming the small plastic bag and sticking it down the front of his pants. "What are they gonna do?" he said. "Arrest me?"

A few minutes later, Eddie, Spanky, and I were taken out to a central area where there was a metal detector and a desk manned by an officer who uncuffed us and had us empty out our pockets into a tray. Out came wallets, keys, loose change. Before going through the metal detector I had to take off my shoes and bang them together, then pull up my pant legs and roll down my socks. I was thinking about Eddie's Jockey shorts stash, wondering what they'd do to him if they found it.

Our next stop was a much larger holding pen with steel bars and low metal benches running the length of two of the pale yellow tile walls. There were about fifteen people inside—all bookie crews from the sweep. "Well, this beats the shit out of the other place," Eddie said as we took seats next to Bernie, Bob, Sal, and Michael on the cold metal bench. We were no longer cuffed.

"It's fuckin' paradise," Bob said, looking up. "Say hello to our friends the roaches."

Shiny bullet-sized roaches were scurrying across the walls and along the floors. There was a toilet in the cell, against one wall, lidless, seatless, and toilet paper-less.

Over the course of the next few hours the cell filled up with other prisoners, mostly young black guys in hooded sweatshirts and unlaced sneakers, some of them slouchy and sullen, others loud and obnoxious. One head-bopper in cornrows puked in the vicinity of the toilet, then lay right down on the floor of the cell and went to sleep. For me, sleep was impossible. There was too much noise, too much activity. When I leaned my head back against the roach-infested wall and closed my eyes, I felt too exposed. Bernie, who had been busted ten times before, said we shouldn't count on being out before the next day—at least another six hours. He was having a hard time breathing, poor guy. They'd taken his Benzedrine inhaler from him.

Even though we couldn't sleep, no one felt like talking; we were too tired and depressed. I was also hungry. I'd made the mistake of skipping lunch. At six in the morning a guard opened the cell door and dropped a box of sandwiches on the floor outside. Stepping over bodies, several of us went to investigate. The box contained cheese sandwiches—stale white bread surrounding one slice of processed cheese —and little cartons of milk. Wonderful for us lactose-intolerant Jews. Even though the sandwiches were in plastic bags, the sense of ambient filth was so overpowering that I expected to see bugs inside. I checked carefully before taking a bite.

"Don't eat too much, Pete," Bernie said to me. "You don't want to have to take a shit."

He was right. Better to go hungry than to have to use the toilet. I took another bite and stopped there.

At 7:30 a.m. they started calling names. By 8:30 nearly all the bookies, including my whole crew, had been moved out. Bob was the last one called before me, and as he got up to leave, he squeezed my shoulder. "I'm sure you'll get called next, Pete. But if you don't.. . it's been nice knowing you."

"Thanks a lot," I said.

The cell door was locked behind him. I put my hands on my forehead in visor fashion and stared at the floor, waiting.

It couldn't have been more than ten minutes later that my name was called, but it was quite possibly the loneliest ten minutes I have ever spent. When my name was called I felt as if I had won the lottery.

I was taken up a flight of caged stairs to a darker and gloomier row of cells, all of them packed like subways cars at rush hour. A guard who was stationed at a desk got up from the novel he was reading— *The Executioner No. 43*. On the wall beside the desk was a poster with a tiny photograph of a black man wanted for homicide, next to which someone had drawn an arrow and written "actual size of head." The guard led me down the row of cells. I saw Sergio Tacchini first, leaning up against the bars, smoking a cigarette. To my relief, my whole crew was crowded inside behind him, a couple of them having secured bench spaces, the rest standing. Everything was relative, but even Bernie and Eddie looked like a class act in here. It was hard to believe I had ever thought of them as lowlifes.

Bob clapped me on the back. "Good to see you, Curly. Thought you might have found yourself a boyfriend downstairs."

"Go to hell."

"We're already there," Bernie said.

We were twenty-five guys in a cell measuring roughly eight by fourteen. The adjacent cell was just as crowded. But the makeup of the two cells was very different. Most of the men in our cell were whites (the bookies), Hispanics, and light-skinned blacks; our neighbors were all black. There was no question that we had been segregated. To my shame I found that I was glad of this.

It was not much to be glad of. This cell was even dirtier than the one downstairs had been; I could feel the filth in a visceral way, the way you feel germs on a sick ward. The open toilet was encrusted with shit, the bowl itself clogged with feces and old cheese sandwiches. And though the smell was suffocating, after a while it was just absorbed into the general feeling of claustrophobia.

There was room enough on the two benches for nine people to sit. Bernie and Michael had managed to get seats, and were nice enough to trade off with the rest of us every so often. But the wooden benches were no great treat. They were narrow and hard, and if you tried to lean back, the bars of the adjoining cell dug into your back. The best position was elbows on knees, hands on forehead—a posture of utter despondency.

Still, I tried to sleep. I closed my eyes and drifted into a haze of exhausted boredom. After what seemed like hours I would open my eyes, ask someone the time, and find that only a minute had passed. It was impossible. Unthinkable. Time couldn't possibly move so slowly.

At 10:30 a.m. Sergio Tacchini's lawyer appeared, and Sergio and his cronies talked to him through the bars, the rest of us straining to hear what was being said.

What we heard was that the ballpark on getting out was 7:00 to 9:30 that night. There was also a ten percent chance of nothing happening until the following day.

Ten percent chance? That was much too big! Where was our lawyer, goddammit? Why hadn't anyone come to see us? What were Steak Knife and Monkey doing for us? The cheap bastards!

In the other cell, the blacks had started chanting rap songs, pounding out the percussion line on the wooden bench, reciting lyrics that they all seemed to know: "Whatcha want, nigger? Whatcha want, nigger? Fuck the world! It's only for the white man!"

I looked at them, revolted and amazed. How could they be singing, here in this disgusting pit, after a night of no sleep? It was as if they could have been anywhere. It didn't matter where they were.

The idea of it not mattering scared me and filled me with grudging admiration. I had been given so much in my life— and had squandered it so casually. I knew the difference between this place and other places. And yet I had ended up here anyway.

Over in the other cell a guy unzipped his jeans and sat his ass down on the shit-encrusted toilet rim. Bob and I traded glances. No way, Jack. Not us.

When the guy flushed, the toilet backed up and overflowed.

"Oh, shit! *Goddamn*, nigger!" someone in the cell yelled.

There were calls for the guard. "Yo, man! Send someone over here with a mop! This ain't no joke! We ain't joking!" But the guard didn't move, just stayed at his desk reading his cheap novel. The dirty toilet water slowly seeped from their cell into ours, spreading across the floor like a disease. I thought of the stories I'd heard of tuberculosis and hepatitis B in the New York jails, of AIDS. The guard's indifference infuriated me.

"So much for the presumption of innocence," I said to Bob. "We're just fucking animals in here. That's how they treat you. You get arrested, you're guilty."

"We *are* guilty," he said.

"That's not my point."

"They's guys at Rikers been waiting months just for their case to come up," said a Hispanic guy who'd been listening in on our conversation.

Someone said, after a few names were called and no one answered, "Well, that's it until at least six o'clock."

I felt a crushing disappointment. I was starving and tired; barely able to keep my head up, unable to put it down. What if we didn't get called at six?

A new guard came on to relieve the old one. There was a candy machine next to the desk that seemed to have been placed there for the express purpose of torturing us. Sergio Tacchini now managed to get the new guard's attention by offering him five bucks for a candy bar.

They struck a deal. The guard let him out of the cell, and for ten bucks he allowed him to feed dollar bills into the machine for a minute.

The black prisoners were up against the bars of their cell, arms outstretched, dollar bills in their hands. "Give the niggers a candy bar!" one of them shouted.

But Sergio's time was up; the guard told him to return to his cell. He came back smirking, his fists full of Oh! Henry and Mounds and Hershey bars.

He doled them out to his crew. Some of the blacks now shifted around, stretching their arms into our cell. Sergio shrugged. "I didn't have time to get no more. This is all I could get," he said.

I felt bad, guilty, though there was no reason for me to; I had gotten no candy bar myself. But I saw the way things were, and felt implicated.

A few of the guys in the other cell continued to wave money at the new guard, and finally he approached them. He was fat-cheeked and had tiny squished eyes, and his neck was one big shaving rash. He took a dollar bill from one of the outstretched hands and went over to the candy machine. There were cries and demands for different brands. Starburst. Milky Way. Snickers. The guard fed the dollar in. He pushed a button. A bag of Whoppers whirred forward and dropped. Instead of taking them to the cell, however, he went back to the desk and sat down.

"Yo, man. Tha's ma dollah! What you doin' wit ma candy? Over here!"

The guard opened the bag, tossed his head back, and poured a few of the malted milk balls down his throat. The protests from the other cell grew louder, angrier.

I shared their outrage. The guard swiveled around in his chair. He poured some Whoppers into his hand, then bent forward and rolled them across the floor. To my amazement, one of the black prisoners actually bent down and scooped up the small chocolate-covered candies that had rolled within reach. He wiped one against his red flannel shirt, then popped it into his mouth.

"Nigger, what you doin'?" one of his cellmates screamed at him.

"Fuck you, man! You think I give a fuck? Fuck that rolling candy-ass motherfucker. I'll eat what I want!"

The screaming continued. I sat down again and buried my face in my hands. There was a dull ache at the back of my throat. I looked at Bernie, who was sitting next to me. He didn't look good at all. His face was ashen. He was struggling to breathe, forcing air into and out of his lungs by puffing out his cheeks like a blowfish.

"Bernie, you all right?"

He nodded, as if the act of speaking would require too much energy —or air.

"Are you sure?"

He nodded again. He was trying to tough it out. He knew if he had to go to the hospital it would delay our arraignment. He had told me about that happening, not with him but with someone else, and how the group from that office had spent an extra day locked up.

"Hang in there, Bernie," I said, patting him on the shoulder. "Just try to hang in."

He nodded, looking straight ahead, puffing out his cheeks.

Michael appeared in front of me. He'd been talking to one of the bookies from another office, and he now reported that a guy in their office who was busted a few months before—his first time, no priors —had ended up doing thirty days at Rikers.

"No," I said. "No way."

"Thirty days. The Brooklyn D.A. is supposed to be a real prick."

I sank deeper as the clock made its way around to six o'clock, minute by tortuous minute. What was it they were always saying? "If you can't do the time, don't do the crime"? I knew now that I could never come back to this place, even for one day. But what about the others? Would thirty hours in lockup be enough to discourage them? It was clear that if real sentences were handed down for bookmaking—sentences of more than a day or two in jail—few of the guys I worked with would want to take the risk. Which raised the question: If the authorities could put us out of business by imposing stiffer sentences, why didn't they? Did the cops need bookies around for safe and easy busts? Did the city need the revenue we provided from the fines and seizures? Did the politicians need something they could use to generate newspaper headlines and give the public the idea that they were really waging a war on crime?

At 6:30 a guard entered the cell row with a clipboard. Everyone in both cells perked to attention. When he called off Bob's and Spanky's names, I felt a rising excitement, and even though no one else got called, it was as if a dark cloud had lifted. I began allowing myself to imagine all the things I had not allowed myself to think of before: food, bath, bed, sleep.

Another hour passed before the rest of us were called. We were lined up outside the cell and cuffed to a long chain. Bernie was right behind me. He looked terrible, all his energy focused on breathing. When a guard gave us the order to move, I told the guy at the front of the line to take it slow. We had an old guy who couldn't go very fast.

Even so, when we reached the second level of stairs, I could see Bernie wasn't going to make it. I stopped walking.

"C'mon, move it," the guard said.

"The old guy's got emphysema," I said. "He can't climb these stairs."

Bernie had already sat down on the stairs where he was. I asked him if he was okay, but he couldn't answer, even with a nod. The guard came over and uncuffed him, then told the rest of us to keep moving.

We were taken to another cell three flights up, at courtroom level, and half an hour later we got to see our lawyer, who made some inquiries and found out that Bernie had been sent to the hospital. Fortunately he managed to get our cases separated, and at nine o'clock we went before a judge. A court date was set. We were released without bail.

Monkey met us in the corridor outside the courtroom with a brown paper bag full of money. He gave us each several hundred dollars. "I hope none of you guys got a dame waiting home for you. Because I don't think there's a chance in hell she'll let you in the front door. You smell like dead rats."

We went out into the fresh frigid air of a January night. I gulped air like I was drinking it.

Back at my apartment—a palace to my eyes—I took off all my clothes and tied them up in a garbage bag, which I set by the door. I got into my bathtub and ran the handheld shower over my head and body at full force. I was so tired and yet so vulnerable to sensation that the simple act of spraying water over my body was like sex. I leaned back against the cool porcelain curve of the tub, and let the warm water wash over my face, shoulders, and chest. I let it run and run and run.

Right before I got into bed I hit the play button on my answering machine. Snugged in between the cool sheets and under the luxurious weight of my comforter, drunk with exhaustion, I listened to the various clickings and whirrings. My eyes were closed, and it was only faintly that I heard my mother, my friend Ezra, and then a voice, strange at first until I realized it was my own, saying not to expect me home soon.

It seemed a lifetime ago that I had made that call. Images came to me: the guard rolling the Whoppers toward the other cell and the way the one guy had rubbed them clean against his flannel shirt; the clear, surprisingly untroubled face of a twenty-five-year-old who had beaten up his girlfriend and whom, in a ludicrous gesture, considering my own situation, I had advised to seek psychiatric help; the yellowish eyes of the bookie in the Sergio Tacchini warm-ups; the way Bernie had

looked, sitting down on the stairs when we were so close to getting out
....

The fatigue in my bones was delicious; I had a sense of the energy that would be mine after sleep. No doubt it would be mixed with anxiety—my old worries about money and what to do with my life. But before I could dwell on that too long, sleep overtook me. In my dreams I went somewhere far away, to a place where I was safe and where the sun shone brightly but I didn't have to squint.

Epilogue

Two months later, at the beginning of March, our case went to court.

Bob was the only one I had seen after the arraignment, although I'd kept up to date on the others. Since the bust, things had been crazy. Bernie had suffered a heart attack and undergone quadruple bypass surgery; Monkey had decided he didn't need the aggravation of the business anymore and had taken his pregnant wife and his son and moved to Phoenix, where he was supposedly going to open a "titty bar"; Eddie had disappeared for a couple of weeks, then turned up at a hospital where tests revealed that he was HIV positive; and Pat, convinced that we had invented the whole bust story just to get rid of him, had hooked on with another office.

Our legal problems had been more serious than we expected. It turned out that one of our players, a bookie from Brooklyn who made layoff bets with us, was connected to the Gotti family. He'd been the primary target of the Brooklyn D.A.'s investigation and wiretap; we'd just been caught in the web. It was a bad place to be. Charles Hynes was a notoriously tough prosecutor. Gambling laws might be hypocritical when seen in the light of lotteries and OTB, but in Brooklyn, as opposed to Manhattan, they were taken seriously. Especially when a lot of time and money had been invested in a case.

In this instance the prosecution started out asking for felonies on all of us, but after negotiations between our lawyer and the Assistant District Attorney and some moments of genuine concern, a deal was cut. The state agreed, in addition to the levying of stiff fines, to misdemeanors for Eddie and Bernie, who had priors; one felony, no jail time, for someone else, our choice who; and dismissals for the rest of us. No one wanted the felony, of course, so Steak Knife finally had to offer to pay a volunteer. When the price got to twenty grand, Spanky stepped forward. He wouldn't be able to vote in the future or take out a bank loan, but to him, at least, that seemed like a fair trade.

Past the metal detectors, in the vast lobby of the Brooklyn Criminal Court building, I spy Michael, Bob, and Spanky, standing around a square-topped plastic garbage can. Michael and Bob, like me, are wearing topcoats over suits; Spanky is wearing a stiff black leather motorcycle jacket and jeans.

"Hey, douche bag," Spanky says.

I join the group, nodding at each of them, grinning. It's not like it was in the office. Even with Bob there's a slight awkwardness, a formality caused by the gap in time and the setting.

"We're waiting on Eddie?" I ask.

"Who else?" Bob says. "He probably stopped off at some crack house on Avenue D."

We all laugh.

"Don't say anything about the AIDS," Bob says. "I told him I wouldn't mention it to anyone."

"Bernie's not coming?"

"No. They gave him a postponement."

"How's he doing?"

"Much better. He's finally listening to the doctors. Hasn't smoked a cigarette in a couple of months, lost about twenty pounds. You wouldn't recognize him. Did I ever tell you what he said to me right after the heart attack?"

"No."

"I'm talking to him on the phone. He's in the hospital—this is before the surgery. He says, 'Bob, I don't know what happened. Everything was fine. I was sitting there in the kitchen, eating eggs and bacon, drinking my morning coffee, smoking a cigarette ... and boom, from out of nowhere, it hits me!' "

We all share a laugh, and in that moment I remember what I liked about the office, what I've missed most.

We continue to catch up on one another, standing there in the halls of justice. Michael has enrolled in a computer course at NYU; Spanky's been banging nails; Bob has gone out on a couple of job interviews, but I can tell they're unsettled, haven't accommodated themselves yet to moving on.

I tell them that I've started writing again, which is true, although I don't tell them much more. I don't tell them, for example, how I've calculated that the eight thousand bucks in cash I've managed to stash will buy me nearly six months of time. Or that I have pledged to myself that this time I will not self-destruct; I will not be distracted or stop writing halfway through.

I don't tell them about Anna, either. About how in the aftermath of the bust, in asking myself some hard questions, it dawned on me that what I'd been trying to get from her was impossible. No one but me could give me what I wanted. Not her. Not anyone.

Last week I sent her a letter to say that I wasn't mad at her for the way stuff had played out with us. Yeah, she could have handled things better. We both could have. I don't know how she'll respond, or even if she will. I'm not sure I want her to.

Still, I can't imagine having no contact at all, not knowing what becomes of her or finding out five years from now that she's married and living someplace like Seattle. I know I ought to accept the fact that not everything sticks or adds up to something, but I can't get over the feeling that there's a reason we've hung on to each other for so long, and that it goes beyond neediness or fear. I can't get over the feeling that someday everything will come clear to us and then all the struggle, all the pain, will have been worth it.

Stubborn or stupid, I still don't know.

A week after the case is settled, I meet Bob for a drink. They're opening the office again. Steak Knife's made the decision. Michael's coming back. So are Eddie and Spanky. Even Bernie.

"What do you say, Curly? Can I count on you for next weekend?"

"I don't think so."

"C'mon," Bob says. "What are you going to do instead?"

"I'm doing it."

"What, writing?"

"Does it matter?"

"Look, you're coming in. I'm marking you down on the schedule, motherfucker."

"You'll just be wasting ink, Bob. I'm telling you."

"I'm telling you. Sooner or later, one hundred percent, you're gonna come back."

"I don't think so," I say. "I really don't."

"I'll bet you," he says. "You wanna bet?"

I am amused by his persistence. "Sure," I say. "I'll bet."

"Go ahead," he says. "Name it."

"Whatever you want, Bob. Your call."

He looks at me now with narrowed eyes. Is he being hustled? Am I playing the smart side for once? Can he really be wrong about me?

"What's the line?" he asks suspiciously.

"What, you want odds, too?"

"You don't think I'm going to bet you without an edge, do you?"

"No."

"So what is it? What's the line?"

I start to calculate it, but then I catch myself. "Hey, you're the bookie," I say. "You make it up."

AFTERWORD

To say the world is different now than it was when this book was first published twenty-five years ago is to state the obvious. And yet as I sit here sheltering in place in the midst of the Covid-19 pandemic crisis of 2020 and with civil unrest brewing, what strikes me is just how quickly things can change.

When I wrote *Confessions* at the start of the 1990s, it was the early days of cell phones (so early in fact that in the book they're called *cellulars*). For the bookie business, this new technology along with personal computers and the internet led to a sea change in the way things were done. The boiler room offices that we worked in, the elaborate landline setups with clerks crammed into sweaty smoky windowless rooms—those don't exist anymore. When I left the biz 25 years ago, the culture clash between young upstarts like Bob, Michael and Spanky and the old guard of Monkey, Jigsaw and Eddie was a harbinger of the coming revolution in which digital technology would eventually transform not only the bookie world but everything else too.

The story I tell in this book ends with us getting arrested and the aftermath of that, including discussions about how we would do things differently in the future. I remember Spanky and Bob talking about buying an RV and driving around the city using "cellulars" and call-forwarding to evade the authorities. In fact, what happened was that in the year after the bust, Jigsaw and Monkey moved the office to Margarita Island off the coast of Venezuela, using 800 numbers to conduct business (phone calls still being integral to how bets were placed), with runners back in the States handling the pays and collects. Although

I wasn't part of it, the island office was in operation for several years, with most of the guys shuttling back and forth between Margarita Island and New York. It was hugely successful for a time, growing from the 10-man crew we'd had before the bust to a large-scale enterprise that eventually employed 170 clerks, most of them Venezuelan locals. For a number of reasons, it didn't last. First, Monkey dropped dead of a heart attack (at his funeral, the rabbi said about him, "He loved sports and the uncertainty of their outcome"), then Bernie died, then Eddie, then Pat, and eventually the remaining guys, including Michael and Bob either got homesick or were forced to leave Venezuela after Chavez took power and exiled all expats. Back in the States, Spanky continued to run a sheet while he trained as a chef; Michael married a high-powered businesswoman and became a stay-at-home dad; and Bob finally tied the knot with Stacy and had two kids with her, putting his bookmaking windfall to good use by starting a successful biomedical company. Of all the guys still kicking, he's the one I've stayed in touch with. He recently sold his company and said to me afterwards, "Compared to Wall Street, the bookie business is as straight and honorable as it comes."

I thought at the time that the publication of *Confessions* would be all the affirmation I needed to propel my writing career to the next level. And while writing did get somewhat easier for me, it's never gotten easy enough, my relationship to it not unlike my relationship to Anna —equal parts passion and pain. To find relief from the oppressiveness of the blank page and to supplement my income, I play poker a couple of nights a week. Most summers I spend a week in Vegas at the World Series of Poker. This balance seems to be working because I've managed to publish five more books and write several screenplays over the years while finishing in the money at the World Series of Poker a total of nine times. Along the way, I got married and became a father. For the longest time, I didn't think I'd be able to make that kind of commitment to anyone—and yet my wife Alice and I have now been together for over 20 years. We've had our moments of drama, but mostly we help one another stay sane. Nothing has contributed more to our happiness than our daughter, Eden.

As for Anna, she married a former high school classmate after re-connecting with him at their reunion. Her son Nathaniel is a grown man now, in his thirties, and married as well. She and I still talk once or twice a year. Her life these days revolves heavily around dogs, some-thing I never would have envisioned when we were together. She trains dogs and travels to competitions with them, agility stuff mainly, ob-stacle courses that the dog runs through while she runs alongside, di-recting. I don't fully understand it, but then I suppose a lot of people don't really get my poker or sports betting either.

With the advent of smart phones, actual conversations and relation-ships between gamblers and bookies have essentially become unnec-essary. Everything's digital now. You can already bet legally in half the states in the country. It seems only a matter of time before sports betting becomes legal everywhere. Naturally, the major sports leagues have recognized this and instead of resisting the new reality are figuring out how to best cash in.

The Vig took place in a watershed time, a fulcrum moment, the col-lision of the past and future. As with most things that were once illegal and have gone legit, the new paradigm has opened up room for a black or gray market aimed at those who don't want to pay taxes and who'd prefer to play on credit. The bookie business will never return to being the kind of backroom operation that I wrote about, but on a smaller scale, you'll be able to find a guy you can call up like in the real old days, only in this case he won't be hanging out by a hotel lobby phone booth, he'll wherever he is, because he's got a cell phone and he's mobile, and you won't actually have to talk to him, you'll just text him and he'll text you back. And that's the way it'll work, you're both beating the sys-tem, avoiding taxes and government oversight. You Venmo him when you lose, he Venmo's you when you win. It's more personal in a way and harkens back to an earlier era—like candles and paper books and letters. Quaint but useful. It's hard not to think that going backwards like that might actually be a kind of progress—especially during these dark days.

Peter Alson
New York City
May, 2020

About the Author

Peter Alson is the author of the highly acclaimed memoirs *Confessions of an Ivy League Bookie* (Crown, 1996), retitled *The Vig* (Arbitrary Press, 2020) and *Take Me to the River* (Atria, 2006); and coauthor of *One of a Kind* (Atria, 2005), a biography of poker champion Stuey Ungar, and *Atlas* (Ecco, 2005), the autobiography of boxing trainer and commentator Teddy Atlas. His fiction includes a novel, *The Only Way To Play It* (Arbitrary Press, 2020) and a story in the collection *He Played For His Wife* (Simon and Schuster, 2018).

His articles have appeared in numerous national magazines, including *Esquire*, *Rolling Stone* and *The New York Times*. His article for *Playboy*, entitled "Speed Seduction," was optioned by Arnold Kopelson and Twentieth Century Fox. The Ungar biography, *One of a Kind*, was optioned by Graham King and Initial Entertainment. And is currently being developed by 3BlackDot. *Confessions of an Ivy League Bookie* was originally optioned by Fred Zollo and Castle Rock Pictures, later optioned by Mace Neufeld and Paramount Pictures, for whom Alson wrote the screen adaptation. His TV pilot, *Nicky's Game*, starring John Ventimiglia and Burt Young, was produced by LaSalle-Holland and appeared in the New York Television Festival and the Vail Film Festival.

Alson lives in New York with his wife, screen and television writer Alice O'Neill, and their daughter, Eden River.